SECRETS

OF

PONDS

AND

LAKES

by

John O. Snow

Guy Gannett Publishing Co.

Portland / Maine

Published by Guy Gannett Publishing Co., 390 Congress Street, Portland, Maine 04101. September, 1982

First edition printed in the United States of America by Gannett Graphics Augusta, Maine 04330. September, 1982

Library of Congress Catalog Card #82-80952

ISBN # 0-930096-30-4

DEDICATION

To Cindy, my forever companion.

ACKNOWLEDGEMENTS

I would like to express my gratitude to those people who have helped in the preparation of this book: to Mark McIntyre who read the manuscript and made helpful suggestions, to Jak Dowling who wrote the foreword, and to Cynthia Snow who typed the original manuscript and provided invaluable assistance.

CREDITS

Photo: Leonard Lee Rue III,
and the Author
Art: John O. Snow

FOREWORD

by Jak Dowling

"Thoroughly researched" is inadequate praise for *Secrets of Ponds and Lakes*, John O. Snow's second book. Its beginnings, like those of his earlier *Secrets of a Salt Marsh*, date back to his childhood.

Snow grew up and still lives close by Maine's Scarborough Marsh. His fascination with the marsh and its life grew from fishing trips there with his father. A freshwater aquarium, a gift on his tenth birthday, triggered his interest in ponds and lakes.

Snow never faced the problem of "finding himself." A degree in biology from the University of Maine at Orono, and teaching life science at his hometown junior high school, were predictable outgrowths of his childhood interests and his environment.

Both "Secrets" books have all the value of formal textbooks plus the fascination of a gripping novel. The author's technique for imparting information recalls to the analytical reader the methods of the tribal storyteller of early civilizations. It was his duty to pass on to younger generations the history and accumulated knowledge of the tribe.

Each of Snow's books constitutes a complete text on its subject, equally spellbinding for adults and youngsters. His "storyteller" approach eliminates, without loss of value, the stretches of boredom so often found in formal textbooks.

Secrets of Ponds and Lakes brings to the reader a sense of wonder at the profusion, complexity, and interdependence of the plants, animal, and aquatic life in Maine's inland waters.

The author describes the simple equipment needed to gain a beginning appreciation of the normally-unseen life there. A piece of nylon stocking stretched over a hoop to capture tiny life forms and an inexpensive microscope to study them opens the door to the teeming underwater world.

His account of a family camping and canoeing trip illustrates the added pleasure and excitement given to such an outing when one has a broad knowledge of aquatic, bird, and animal life common in and around Maine's ponds and lakes.

Snow combines the enthusiasm of an outdoorsman with the skills and knowledge of a professional biologist and the evangelical zeal of a born teacher to lure his reader into the little-known world literally at our feet.

TABLE OF CONTENTS

A bold mosquito drains blood from a frog, an amphibian which eats many kinds of insects, including mosquitoes.

Secrets To Solve

For my tenth birthday, I received a large cardboard box with a blue bow taped on top. The lack of typical birthday wrapping was outweighed by a black ink message printed on the side, "Contents: one freshwater aquarium." The steel supports secured the flat plates of glass for distortion-free viewing, unlike the motley collection of jelly jars assembled on my window sill. Those jars housed a precious collection of curious plants and animals scooped from a neighbor's pond. Tadpoles, diving beetles, snails, minnows, and water weeds were moved to the new aquarium to live out a portion of their lives before my eager eyes. By day, green plants carried out their magical process of sugar production from sunlight, water, and carbon dioxide. Tiny bubbles of oxygen clung to their leaves whenever sunlight struck the plants. This oxygen dissolved in the water to provide the breath of life for the animals. Tadpoles scraped algae from the stems, and minnows gobbled mosquito wrigglers with gourmet greed. At night, the giant water bug stabbed a tadpole with its straw-like mouth and drained the life juices. These life and death struggles were a daily part of my ten-gallon community.

Even today, a special sense of wonder and excitement touches me when I explore a new lake or pond. The dip of a pail among the weeds always dredges up a new insect or unusual plant. My curiosity is aroused to discover what animal left the strings of gelatinous eggs draped over the water weed, or why painted turtles spend long hours in the hot sun, or how a slender-legged water strider can walk on water.

Mysteries such as these present a challenge to the curious observer, and reward those who take up the challenge of solving Nature's puzzles. Patience, careful observation, research, and a desire to find the answer frequently yield the solution. Often, however, the freshly acquired knowledge leads to greater mysteries, for the profusion of life in lakes and ponds is a dynamic fabric of interwoven threads, each dependent upon the others.

During your investigations, concentrate most of your efforts on one lake or pond. This way, you become familiar with the entire body of water and more detailed aspects can be observed. A short pause at several ponds reveals only the obvious.

Research the history of your pond or lake. Try to discover how it formed. How old is it? What changes have man's activities caused?

Investigate the physical qualities of your subject. What is the temperature of the water near the surface and close to the bottom? How does the temperature vary with the seasons? Is the water acid or alkaline? How deep is the water?

Observe the plants and animals struggling to survive. Which plants grow close to shore or deep in the water? How are floating plants adapted to the surface? Which animals live buried in the sediments or attached to the plant stalks?

Interesting adventures are everywhere in the world of ponds and lakes. Your curiosity is the key to unlocking the secrets of this freshwater environment. Explore with your senses. Listen for the deep-throated call of the bullfrog and the "*kong-ka-ree*" of a red-winged blackbird. Watch the graceful curve of a water snake gliding across the pond's glassy surface. Breathe in the fragrance of spruce and pine.

As your appreciation for this habitat grows, binoculars, guide books, magnifying glass, and camera will expand your awareness. If a microscope is available, an entirely new world of glass-encased diatoms, sparkling desmids, and flowing amoeba can be revealed.

Enjoyment of any environment is increased by comfort. This is also true of ponds and lakes. Appropriate clothing will depend upon the time of year and the ferocity of the insects. Old sneakers are useful for climbing on rocks (Watch out for wet, slippery rocks!) and wading in shallow water. Check your feet and ankles occasionally. Leeches enjoy a free ride attached to your skin, but they will enjoy a free meal even more.

Not everyone expresses their appreciation and enchantment with fresh water in the same way. Some sit quietly for hours observing life and death struggles. Others relax in the soothing moods only water can create. Still others come to enjoy the beauty but irresponsibly spoil and vandalize the environment. Young lovers who proclaim their undying devotion by carving initials in birch trees or by painting hearts on glacial boulders visibly reduce Nature's artistry. The casual picnicker who dines at the lakeshore and leaves behind a corn chip bag, empty soda can, and fried chicken bones wrapped in aluminum foil has learned very little from his communion with Nature. An appropriate sign posted at a national wildlife refuge reads, "Since no entrance fee is charged, consider your fee to be one bag of litter removed from the refuge." That price is certainly affordable by everyone.

This book deals with ponds and lakes in my native Maine and how I have enjoyed them. Although there are physical and biological differences between these bodies of fresh water, the information contained here will apply to most ponds and lakes where the water is shallow enough for plants to grow from the bottom. I hope this book will guide you in the enjoyment of discovery and the appreciation of God's world.

John O. Snow
Scarborough, Maine

My children, Jill and Matthew, take time to enjoy a mountain lake.

Birth And Death Of A Lake

Over fifty thousand years ago, the earth's climate cooled. At first the change was minute. The effects were nearly imperceptible. Perhaps scattered patches of snow lingered in the shade of rocks and shrubs throughout a Canadian summer. The following year, a few more inches of snow remained. More patches persisted. Centuries passed. Unmelted snow accumulated to nearly two miles thick. Sluggishly, the ice pushed southward as the Laurentide Glacier of the last ice age.

This immense sheet of ice crept over the northeast. Other glaciers blanketed much of North America and Europe. The sculpturing force of the flowing ice chiseled jagged mountain peaks into rounded domes. Broad valleys, deep furrows and irregular pits were gouged from the earth's crust. Stones were crushed into pebbles, and pebbles were ground into sand.

Great piles of eroded rock and soil pushed ahead of the advancing ice sheet until the climate warmed about eleven thousand years ago. Then, the warming weather melted the glaciers and forced their retreat toward the north, leaving behind mounds of debris known as terminal moraines. These mark the glaciers' southward advance.

Terminal moraines, although greatly weathered over the centuries, can still be found today from New England to the Rocky Mountains. They extend across the continent as a gently sloping ridge. Long Island, Block Island, Martha's Vineyard, Cape Cod, and Nantucket are examples of glacial moraines.

As the glaciers melted, torrents of frigid water gushed from beneath the melting ice into countless churning streams. These freshets rushed across the glacial plain to join larger streams and then rivers. The seaward journey of numerous streams and rivers was blocked by moraines. These natural dams flooded large areas, creating narrow, finger-like lakes. Thousands of irregular dents and scrapes filled with water and splotched the land with a confusion of ponds. Some were shallow pans which

Bigelow Mountain reflects in Flagstaff Lake, a large man-made lake in Western Maine.

would disappear in a few years. Others were small, but deep, kettle lakes. Although devoid of living plants and animals, a new freshwater environment had been born.

The remolded earth of eleven thousand years ago must have resembled our arctic tundra. It was a bleak, bitter world swept by biting winds. No trees or shrubs grew. Animals forced southward by the glacial sheets had not yet returned.

Water became a marvelous instrument of nature. Rains pattered and splashed against the land, returning a vital ingredient to the soil. Springs bubbled forth from subterranean flows to feed the freshets. Murmuring streams moved massive volumes of silt and sifted out the particles according to weight. Rivers constructed deltas from these fragments. Crashing ocean waves abraded the strongest rocks into flecks of silica. Water molecules dissolved minerals, salts, and gases.

Yet, water's most essential quality lay not in its capacity to alter landscapes or erode rocks. Water was, and is, the vital fluid which makes possible the existence of life.

Millions of years of evolution had molded the cells which would inherit the glacier's legacy. Borne by winds and wandering birds, protective spores, eggs, and cysts invaded the lakes. They required little to survive: sunlight, minerals, dissolved nutrients, and gases. Finding ready supplies of these ingredients, tiny plants flourished. Through the unique process of photosynthesis, the green plankton converted the energy of sunlight into sugar. This sugar became the food supply of the plants. It energized their growth, repair, and reproduction.

The microscopic plant cells divided rapidly. The blue water was tinted with the greens and yellows of plankton blooms. Minute animals grazing on these plants digested the sugar-rich protoplasm to supply their own energy needs.

As generations of plankton passed, each tiny cell added a speck of organic matter to the lake bottom. Bacteria recycled these remains into fertilizing nutrients. Blended with the silt washed in daily by inflowing streams, a rich sediment accumulated.

Wind-borne seeds also drifted to the lakes' shores. Grasses, ferns, and wildflowers sprouted through the carpet of sphagnum moss already surrounding the lake. Trees and shrubs stretched toward the sun.

Attracted by the green shoots and floating algae, migrating ducks and geese settled onto the lake. The night before, they had rested in a southern marsh and their feet were still caked with mud. As they swam and dabbled, the mud washed from the webbed feet. Hidden in this mud were eggs of worms, snails, and even fish. Seeds of cattails and rushes clinging to their feathers were washed away. Within days, the eggs hatched and the seeds germinated. The new cargo of life grew in the lake.

Migrating Canada geese transport many seeds, spores, and eggs on their webbed feet.

The scented water lily displays its large fragrant blossom during the morning hours only.

The plants spread rapidly and extended their range further into the lake. Broad-leaved lily pads, tethered to the bottom by slender stems, blossomed with large white blooms. Yellow spatterdock and purple pickerel weed flowered above the surface. Ferried by their own wings, insects investigated the blend of scents near the lake. Robber flies hunted above the water's surface while diving beetles snatched prey beneath. Worms burrowed through the mud. Clams siphoned plankton from the water. Sunfish and sticklebacks built nests among the roots and stems which forested the lake's bottom. In the open water, diatoms and desmids collected sunlight like tiny greenhouses.

Time passed and the lake matured. A thick ooze, enriched by large amounts of dead plant matter, accumulated along the bottom. This natural compost nourished many new plants which succeeded the original pioneers. Large mats of floating weeds stretched into the lake. Their roots and stems collected silt from inflowing streams. Combined with soil washed in by rains and decaying plant debris, the edges of the lake slowly filled. Black-masked raccoons scurried across these supporting mats of vegetation in search of careless crayfish and frogs. Water squeezed through the raccoon's toes, but they did not sink. A startled frog leaped away from the inquiring nose of the raccoon. A quick dive, followed by a few short strokes, buried the amphibian in the thick sediments. Each century added about one foot of material to the lake's bottom; thus, as the years passed, the depth diminished and more open water disappeared.

During its mature years, the lake harbored a multitude of plants and animals. Many migrating waterfowl depended upon its bounty of food. Visiting mammals foraged and drank at its shores. Turtles and water snakes basked on rocks and floating logs. Insects, mussels, snails, crayfish, worms, protozoans, and fish all lived in the lake at some time. As conditions changed, certain populations died away to be replaced by more suitable ones.

Eventually the lake shallowed. Plants growing from any point on the bottom could reach the surface. Very little temperature change occurred from top to bottom. The lake had always had two distinct temperature layers during the summer and winter. Additionally, the area of open surface water diminished to a few acres. Thousands of years had passed and the lake was dying. In its old age, the lake had become a pond.

Then, the aging continued, but at a faster rate. More plants could now reach the surface. They yielded large quantities of organic matter to fertilize the bottom and thicken the sediments. Once more, different populations of creatures inhabited the fresh water, particularly those able to endure warm temperatures and low amounts of oxygen. Floating algae and duckweed covered the surface. Smartgrass and arrowhead grew where trout once swam. Minnows, shiners, and bullheads replaced the bass and pickerel. Cattails, rushes, and sedges made a final assault on the open water. The pond was near death.

During the next century, the pond became a marsh. Only puddles of water remained, surrounded by squishy land. Trees and shrubs grew in place of lily pads. The swampy land was ideal for the spring peepers, salamanders, and toads which laid their gelatinous eggs in the deeper puddles. Worms and crayfish still lived in the spongy soil, but the clams, turtles, and muskrats had disappeared. Plankton flourished in the puddles, but their period of growth was limited. The warm days of July and August evaporated the remaining water. The microbes formed resistant spores and cysts to protect them until the water returned in the fall.

Finally the pond was gone. For thousands of years it had nourished and sheltered a wealth of life in its water-filled basin. Many changes had taken place since its glacial birth and those changes were not yet finished. Filled with a nutrient-rich soil of decaying plants and animals, the lake basin now harbored a flat meadow. In the years to follow this meadow would be home to many land creatures. Eventually the meadow would be invaded by red maple, and white pine to produce a forest where a glacier once gouged out a section of the earth's crust fifty thousand years ago.

Ponds are born in other ways too. When fast-flowing streams reach the lowlands, their rapid plunge toward sea is slowed. A river with gentle currents and sweeping curves develops. As erosion extends the meandering loops of the channel, the river can twist back upon itself and cut the loop off from the river. The isolated meander, really a U-shaped pond, is known as an oxbow lake. These "lakes" quickly fill with cattails, pickerel weeds, and water lilies, a clue that these bodies of water are not truly lakes. Although the distinction is a general one, ponds are considered small areas of standing water, shallow enough to permit plant growth over most of the bottom. In addition, the temperature of a pond's water remains fairly uniform from top to bottom.

By comparison, the water of a lake is generally deeper and divided into two distinct layers. During the summer, surface water warms under the constant radiation of the sun. Deeper in the lake lies a layer of cold water. A region of abrupt temperature change separates them and prevents any mixing except in the spring and fall. At these times their temperatures vary little from top to bottom and strong winds quickly cause currents to blend the layers. These "overturns" restore oxygen to the deep water while returning nutrients to the surface. This recycling of vital ingredients throughout the lake supplies the needs of both plants and animals.

In shallow lakes and ponds, the water tends to be warm, rich in nutrients, minerals, dissolved gases, and crowded with thick growths of aquatic plants. Beneath the surface, visibility is severely limited by a natural murkiness from an abundance of tiny plants, animals, and rotting vegetation. A large number of fish, including pickerel, bass, perch, chub, and hornpout, commonly inhabit these "rich nourishment" (eutrophic) lakes. Deep, cold water lakes and ponds, referred to as "scanty nourishment" (oligotrophic), are characterized by clear water and rock-strewn bottoms. Very few plants grow from their nutrient-poor sediments or around their shores. Little plankton floats in the water. Trout and salmon reside in the cold water, snatching insects and leeches. A third type, referred to as a "defective nourishment" (dystrophic) lake, lacks adequate drainage. This results in large volumes of organic debris and little decay. Fish are absent from the brown water which acquires an acidic nature from the unrotted vegetation, and the lush growth of acid-producing sphagnum mosses around its borders. In old age, these lakes become the bogs found in the northern coniferous forests.

Obviously, much variety exists among our lakes, ponds, and bogs, not only in the type of life found there, but in the very nature of their physical and historical qualities. As you study your favorite body of fresh water, identify its numerous traits, its possible origin, and its lively inhabitants. Take time to examine each one closely, and you will marvel at the intricate threads which weave the webs of life around each other.

The tiny rosettes of sphagnum leaves account for many of the mosses' unusual traits. The grey-green leaves are only one cell thick and curl to hold water weighing more than the leaf.

Bog Ponds

A velvety carpet of sphagnum moss clings to the hummocks among the black spruce trees. Flickering sunlight sifts through the dark green boughs, splotching the ground with patterns of gold. Spongy moss squishes under foot as water is wrung from tiny leaves. Wood frogs pop into pockets of cold black water collected beneath the exposed tree roots. A few star flowers and bunchberries sprinkle the sphagnum green with their frosty blossoms. Old man's beard, a pale green lichen, the beneficial union of an alga and a fungus, hangs in straggly locks from the lower limbs. Through this twisted maze of dead branches, light reflected from a rippling pond is barely visible. All of these clues indicate we are approaching a unique and rarely visited world—the bog pond.

Pushing through the thicket of spruce branches, the mature trees give way to smaller versions of the same species. Although these spruce may be a century old, their height is less than your shoulder. The bog pond is clearly visible and completely surrounded by a ring of dwarf trees. Sharing the added sunlight at the bog's border, grows the larch or tamarack (*Larix*). Tufts of soft, pale green spills sprout irregularly from the boughs of this conifer. Small rounded cones, their scales opened like the petals of a flower, cluster some of the limbs. By contrast, the black spruce clothes its branches with rich green needles which hug the tree year round.

Tamarack, however, yields to the approach of winter by fading from green to yellow. The dying leaves fall from the tree, leaving stark sentinels until spring's warmth brings forth new life.

Before the dwarfed conifers, lies a band of shrubs and mosses which quake with every footstep. Even the small trees rock as tremors ripple through the floating mat. Water oozes from the thick cushion of mosses beneath your boots as your feet slowly sink into the spongy plants.

Ahead lies the open water of the bog pond, all that remains of an ancient lake. In ages past, the scouring ice sheets of the Pleistocene era either scraped out the ancestral lake bed or left behind a great chunk of ice buried in the earth. Melting ice filled the basin. Thus, bogs share their origin with familiar bodies of fresh water.

Over a span of many centuries, a natural filling raises the bottom of lakes and ponds as dead plants and sediments accumulate. This same process is at work in the bog pond. Bogs, too, shrink from the bottom up, but because of their poor drainage, they also diminish from the top down. The stagnated water allows sphagnum moss to "creep" over the surface of the water. The profusion of sedges, reeds, cotton grass, and marsh cinquefoil growing along the shore provides a foothold for the sphagnums to spread their gray-green leaves onto the water. As these mosses die, leaves and stems drifting to the bottom, pile up in layers of partly decayed peat. Others collect around the bases of the plants. The lack of drainage hinders the activities of decay-causing bacteria, and the water-soaked peat accumulates with little rotting.

Acids building up in the water help the sphagnums to expand their range while inhibiting the growth of many competing plants. As the peat layers become denser, bogbeans take root and bind the matter together. The thick roots of bogbean also sprout a type of buoyant stalk. Shoots from these floating stalks weave a loose net. Sphagnums cling to this mesh, fill in the spaces, and thicken the floating mat. In May, the bogbean (*Menyanthes trifoliatus*) spreads a cluster of white blossoms above the mosses. Each bloom has five arching petals fringed with long white hairs.

The slow rate of bog decay releases few nutrients for plant growth. Yet, numerous plants of the heath family (bog rosemary, bog laurel, Labrador tea, and high bush blueberries) and the orchid family (rose pogonia and lady's slippers) thrive on the meager supplies of nitrogen and phosphorus, but not without help. Intertwined among their roots lives a soil fungus capable of freeing the necessities of plant life from the acidic peat. In exchange for their rations of nitrogen, the heaths and orchids furnish a cozy home for the fungus in and around their roots.

The fungus-orchid symbiosis succeeds. Numerous orchids paint dabbles of color among the mounds of sphagnums and heaths. Yellow and pink lady's slippers secrete their showy blooms in shaded retreats while other orchids seek full sun. The delicate pink blossoms of rose pogonia (*Pogonia ophioglossoides*), grass pink (*Calopogon pulchellus*), and arethusa (*Arethusa bul-*

Rose pogonia (L) and cotton grass (R) both thrive in the acid conditions found in bogs.

bosa) flower in early spring to be replaced by the fringed orchids as summer begins.

The cup-shaped buds of Labrador tea (*Ledum groenlandicum*) open into waxy-white blooms during May. These are accompanied by the pale pink flowers of its close relative, the bog laurel (*Kalmia polifolia*). Delicate bell-shaped tresses dangle in clusters from the sprigs of leatherleaf (*Chamaedaphne calyculata*). This bog evergreen bears numerous upturned leaves which blaze scarlet during October.

With the arrival of autumn, the inconspicuous cranberry vines (*Vaccinium macrocarpon*) adorn the bog with drops of crimson. The large tart berries, a fruit of a tiny pink blossom, appear.

Tucked among the sphagnum hummocks and on the quaking mat itself grow two unusual plants—the pitcher plant and the sundew. Like most bog vegetation, their need for nitrogen and phosphorus requires special adaptations. Instead of absorbing their nutrients through a tangle of roots, the pitcher plant and the sundew grow specialized leaves to capture and digest the nitrogen-rich bodies of insects.

The red and green, open-mouthed traps of the common pitcher plant (*Sarracenia purpurea*) invite insects with their flower-like appearance. Nectar glands on the pitcher's edge release an enticing honey scent. Lured to the pitfall by these deceptions, the insect finds a convenient "landing pad" at the lip of the pitcher. Inward-pointing hairs direct the insect toward the watery snare. On the sloping walls of the trap, stiff hairs point downward, preventing the prey from any retreat. To guarantee success, a portion of the steep wall lacks the guiding hairs. Caught unaware, the insect slips on the smooth surface and tumbles into the water-filled pitcher. The struggling insect quickly drowns and joins the decaying bodies of spiders and ants at the pitcher's bottom. Here, the nutrient broth is absorbed by special cells to nourish the growing plant.

Numerous, open-mouthed pitchers invite insects to visit their red and green traps.

One insect, however, is not deceived by the pitcher's passive trap. A bog mosquito (*Wyeomyua smithii*) actively seeks the water-filled cisterns and lays her eggs within the pitfall. Wriggling larvae hatch in the trapped rainwater where they feed and grow. Following metamorphosis, the adult mosquitoes pull from their pupal skins and fly free of the open-mouthed traps.

Unlike the obvious pitcher plant, the inconspicuous sundew (*Drosera*) attracts little attention. Colonies of these small plants grow on logs and soil. Some nestle among the mosses. The red, spoon-shaped leaves are adorned with a fringe of long hairs. Each hair bears a glistening drop of sticky "sundew" atop its summit. The twinkling bits of glue lure insects to investigate more closely. An unwary fly or mosquito landing on the red leaves realizes his mistake too late. Legs are held fast by sticky strands. As the insect struggles to free his entangled legs, wings and body touch the adhesive tentacles securing the prey in a snarl of tacky threads.

The insect's weight triggers the tentacles to slowly curl against the prey, and the leaf gradually bends around it. Soon, digestive juices flow from the surface of the leaf to dissolve the victim. Nutrients are then absorbed. Several days later, the trap reopens. Fresh drops of sundew sparkle at the ends of the reddish tentacles waiting to ensnare another meal.

Like all ponds and lakes, a bog is doomed to die. The remarkable sphagnums which help create a bog by dominating the other plants eventually seal off the open water. In time, shedding leaves and spreading roots will replace the water with peat. As drying begins at the bog's border, a mature forest can invade the once soggy edges. The sphagnums starve under the full shade of the larger trees. When dried, the peat begins to rot, releasing nutrients to fertilize the growth of the invaders. Then one day, the bog is gone, replaced by a forest of towering spruce.

The waxy-white blossoms of Labrador tea begin to open on a dewy May morning.

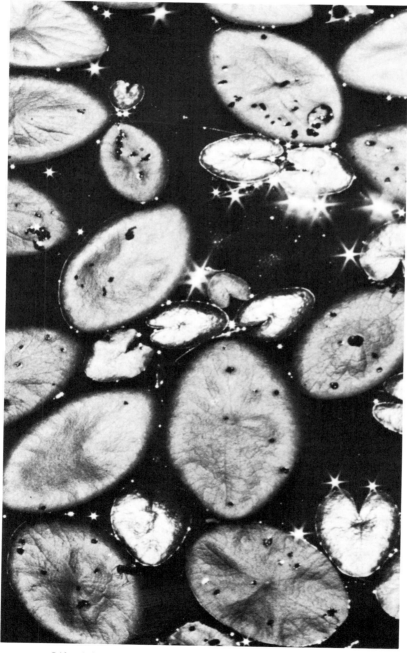

Life-giving sunlight energizes the broad leaves of a lily pad.

Energy And Green Plants

Gazing intently into the shallows, a great blue heron (*Ardea herodias*) searches for the tell-tale movement of a minnow or tadpole. A large bird, the heron stands nearly four feet tall. His stilt-like legs resemble the plant stems among which he wades. Slowly raising his clenched foot from the water, the great blue heron spreads his toes and gently steps forward. Hardly a ripple stirs as he slips his foot beneath the surface. The keen yellow eyes continue surveying the shallow water.

A water strider glides beneath the heron and grasps a mosquito as it begins to emerge from the pupal case. A pond snail topples from the stem of a bulrush where it has been feeding. A wiggle betrays the presence of a tadpole scraping algae from the stem of an arrowhead. Instantly, the heron tenses; his neck coils as he draws his head back. In a blur of motion, the long beak jabs the water and snatches the tadpole from his hiding place. With a flip of his head, the heron swallows the tadpole headfirst. A drink of water follows before he resumes fishing.

Throughout the lake there are similar struggles to survive. The snapping turtle crushes a chub which swims too close to its powerful jaws. A leopard frog flips a sticky tongue from his mouth to catch a blue damselfly. Muskrats tug at the cattails and gnaw the starchy roots, while a pickerel seizes a young bullfrog. Events such as these are easily observed in any pond or lake. Although common, their importance must not be overlooked. Food is the source of energy for all organisms. Growth, repair, movement, and reproduction of cells depend on a continuous inflow of energy. Energy maintains the living body.

The waters of a lake contain most of the ingredients for life: oxygen, carbon dioxide, water, nutrients, and minerals. However, the most important ingredient, energy, is missing. Energy is supplied to the pond by sunlight which falls upon the leaves of green plants in the water and along the shore. As sunlight strikes the leaves, a magical reaction captures the sun's energy and imprisons this genie of life in molecules of sugar. The sugar molecules can then become the basis for the food required by both plants and animals.

This process, called photosynthesis, occurs in all the pond's green plants, from the largest lily pads to the microscopic diatoms. All of these plants contain green chlorophyll within their cells. Chlorophyll is the magical potion which brews carbon dioxide, water, and sunlight into sugar.

These "producers" of sugar are the first link in a chain of life. Animals (herbivores) feeding on the plants use the sugar molecules to energize their activities. The solar nourishment helps them to grow, move, and reproduce.

In turn, meat-eating animals (carnivores) devour the herbivores, and the flesh of the plant eaters transfers its energy to the next link. Eventually, producers, herbivores, and carnivores die. The nutrients locked in their tissues are recycled into the pond by decay-causing organisms (decomposers). Once released, these nutrients enter the sediments and water to fertilize the growth of more green plants.

Examples of food chains are everywhere in a lake. A mayfly nymph eats bits of algae floating in the water. The nymph is swallowed by a small minnow which is gulped by a hungry pickerel. A diving otter may consume the pickerel. When the otter dies, his body is recycled by decomposition.

With careful observation, you can link together many different food chains. These chains can then be interlocked to weave the intricate web of life sustaining the freshwater community.

The most important food-producing plants in a lake seem insignificant. They are not the broad-leafed lily pads or tall cattails but are microscopic, one-celled plants and slimy pond scums. These floating food factories called phytoplankton nourish the entire lake.

To fully observe these miniature plants, you need a collecting net and a microscope. A suitable net for capturing the larger forms of plankton can be constructed from a nylon stocking attached to a stiff hoop. Remove the foot from the stocking and stretch the cut portion over the neck of a jar. Secure this with a piece of string. Water dipped from the pond can be poured through the net to concentrate the plankton in the jar.

Under a microscope's magnification, an enchanting world appears. Crystal diatoms, their living fluids squeezed into overlapping glass shells, twinkle in the light which sustains them. Their sculptured silica walls fit together like a small pillbox composed of bottom and lid.

The glass boxes appear in a confusion of geometric shapes. Each one is etched with fine lines and tiny pits. Inside, yellow-green chlorophyll generates grains of sugar and bubbles of oxygen. Specks of oil scattered through the protoplasm buoy the diatoms near the surface where the sun's light can energize their photosynthesis.

Diatoms, like all plankton, must drift with winds and currents. However, they are not entirely at the mercy of their environment. Pores in the glass cases allow threads of protoplasm to stream out of the box. These strands return through another set of holes. The friction of the protoplasm against the water propels the diatom ahead.

Ponds with disagreeable odors and tastes may have abundant populations of blue-green algae. These simple, primitive cells

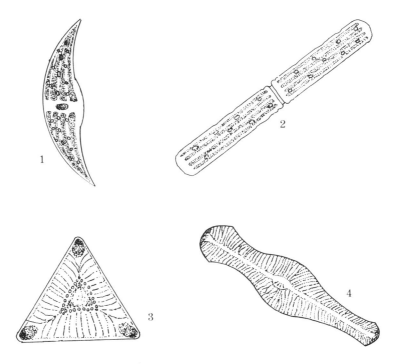

A confusion of shapes and building materials characterize the microscopic planktonic plants. Desmids (1,2) contain a brilliant green chlorophyll. Diatoms (3,4) are glass encased plants.

are plentiful during the spring and summer when their blooms may discolor the water. Under the microscope, many blue-green algae resemble strings of beads.

Perhaps the most beautiful phytoplankton are the desmids. Housed within their transparent walls lies a brilliant green chlorophyll characteristic of the green algae. Each desmid is composed of two halves connected by a clear, narrow bridge. These delicate cells sparkle in the light with a variety of shapes from individual crescents to long graceful filaments.

The green algae include other interesting specimens. *Pediastrum* forms a toothed disc, while *Hydrodictyon* joins into flat, net-like sheets. Its colonies are called the water net. *Spirogyra* forms dense mats on spring ponds. The filaments have a twisted chloroplast (cell part containing the chlorophyll) spiraling through each cell.

Volvox and *Euglena* are exciting organisms to locate because they possess great mobility. *Euglena*, a single cell with both plant and animal features, pulls itself along by lashing the water with a tiny hair-like flagellum. Packed inside its cell membrane are numerous green chloroplasts which manufacture sugar.

Volvox, a giant in this microscopic world, forms a large hollow ball. Thousands of cells are joined into the spherical colony which tumbles through the water. Individual colony members have two flagella, an eyespot, and interconnecting fibers. The flagella, coordinated to beat in unison, roll *Volvox* from place to place in search of food. During reproduction, tiny spheres form inside the hollow parent. Eventually, these young are released through a hole in the parent's wall to begin their own spinning lifestyle.

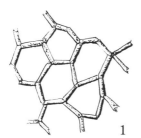 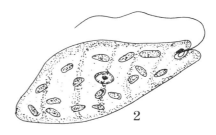

The water net (1) is a series of cells joined into a colony. The *Euglena* (2) is a one celled organism capable of movement with its slender flagellum.

Diatoms, desmids, and blue-green algae have inhabited the aquatic world through their entire history. The essence of their existence was nutured in the ocean two billion years ago. Gradually, they expanded into brackish estuaries, and then to freshwater streams and ponds. Their resemblance to their marine ancestors is obvious.

Other plants evolved to live on dry land, only to re-enter the fresh water of lakes and ponds. These plants developed the familiar roots, leaves, stems, and conducting tissues of land plants. These structures are generally retained by the aquatic varieties, but adapted to fit their new environment.

The scented water lily (*Nymphaea odorata*) possesses several of these adaptations. Thick, flexible stems moor the broad leaves and fragrant white blooms to roots embedded in the mud. Unlike the rigid stem of land plants, the lily's stalk has pockets of air for buoyancy, and the structural supports are greatly reduced. The leaves have a water repellent surface to keep the upper side exposed to the sun. The flattened leaves are ideal for soaking up light to energize sugar production.

Spatterdock blossoms extend above the water on a thick stalk, rather than floating like the scented water lily.

Plants associated with lakes and ponds usually grow in definite patterns or zones.

Emergent plants, like cattails, pickerel weeds, arrowhead, and blue-flag iris are securely rooted into the soil beneath a pond's surface. Leaves and flowers are borne above the water where photosynthesis and pollination can occur. The submerged portion of these plants provide food and shelter for numerous fish, crayfish, worms, and insects. Dragonflies and mayflies regularly pull their nymphal bodies onto the leaves of cattails before shedding their skin and changing into adult insects. Emergent plants may be found as deep as six to seven feet.

Floating plants occupy a second region. Most of these are rooted, like the scented water lily, but some float freely. Duckweed and wolffia are both free-floating plants. Spatterdock, the yellow pond lily, begins the summer as a floating plant, but as the stem lengthens and water levels drop, the leaves and flowers are held above the water as an emergent plant.

Submersed plants like water milfoil, waterweed, and pondweed remain submerged throughout the summer. Although these plants may be rather inconspicuous in the early spring, by mid-summer their dense beds cover large areas. Numerous fish, particularly the young, find security among the thick tangle of stems. These plants are also used for food and nesting material.

Numerous small insects, snails, and fish find the wide pads useful as nurseries, homes, and dining facilities. The water lily leaf-cutter caterpillar (*Paraponyx*) utilizes the pads in several ways. The caterpillar hatches from eggs laid on the underside of the leaf. The hungry larva feeds on the plant while chewing a circle of tissue from the leaf. This piece of plant tissue becomes a tiny green raft which is launched with the caterpillar aboard. When another suitable leaf is found, the larva cuts a second circle and pulls the new piece of tissue over its soft body. The edges are sewed together, forming a caterpillar sandwich. Protected in its new home, the caterpillar nibbles on water lilies and algae as it drifts on the water.

Bright yellow blossoms distinguish spatterdock (*Nuphar sagittifolium*) from all other water lilies. Waxy, yellow sepals surround the true flower in the center. The broad leaves, some twelve inches across, protect the blossom from wind and waves. The thick stems are a cradle for the eggs of many invertebrates, and the starchy roots feed both muskrats and beavers.

Growing in shallower water, the violet-blue flowerspikes of pickerel weed (*Pontederia cordata*) attract many pollinating insects. The ten-inch, arrow-shaped leaves are dark green and shade the water beneath them. Occasionally, slender pickerel lurk among the shadows and stems waiting patiently for unwary fish.

On more sandy bottoms, the white button flowers of pipewort (*Eriocaulon septangulare*) grow from tufts of pointed leaves. Nearby, arrowhead (*Sagittaria latifolia*), named for leaves which resemble an Indian arrowhead, sprouts three to four feet in height. In late summer, it produces whorls of snow-white flowers, each with three petals. Thickened parts of the roots, called tubers, are a valuable food source for waterfowl; hence the name, duck potato.

Still closer to shore, blueflag iris (*Iris versicolor*) blossom during May and June. The three violet-veined sepals are splashed with yellow and white. These surround the purple-blue petals in the center. The long narrow leaves resemble those of the cattails.

Both broad and narrow-leaved cattails (*Typha*) flourish along the shallow edges of lakes and ponds. Each spring the slender leaves poke through the tangled mat of the previous year's growth to reach a height of six to eight feet. The familiar brown "cat's tail" is composed of thousands of tiny flowers. Each one becomes a wind-borne seed after pollination. Marsh wrens use

A blueflag iris blossoms in early spring to bring color and beauty to many ponds.

tufts of these flowers to line their nests which hang among the cattails.

Numerous grasses, sedges, rushes, and bur-reeds flourish in the pond's damp edges, as do the broad leafy fronds of ferns. Their lush growth frequently droops over the water to shade the hiding places of smallmouth bass.

Farther from shore, the truly aquatic pond weeds root to the bottom. During the summer, their dense, submerged growth gives shelter and food to fish, insects, and other animals.

Floating freely on the surface, the tiny green leaves of duck-weed (*Lemna minor*) can blanket an entire pond. This smallest seed plant is a favorite food of waterfowl. Beneath their flat-tened leaves dangles a jungle of rootlets which shelter myriad organisms. Since duckweed rarely produces flowers and seeds, rapid multiplication depends on budding—the growth of new individuals on part of the leaf which then breaks off to become a separate plant.

Most aquatic plants absorb the nutrients required for their existence from the surrounding water. Bladderwort, however, has developed an unusual method of acquiring extra nourish-ment. Sprinkled among its tangled stems hang countless tiny sacs or bladders. Hairs and bristles are arranged on each bladder near a trap-door mouth. When a tiny animal touches one of the hairs, the bladder instantly inflates. The surrounding water and the creature are sucked in. The trap closes, locking the meal inside where it is digested, and its nutrients are slowly absorbed.

As you explore the plant world around your pond, be aware of the obvious, but also search for hidden secrets. Lift a lily pad and examine its under surface. The eggs of insects and snails are frequently laid there. Use a hand lens to examine the rootlets of duckweed for minute crustaceans and *Hydras*. Study the special adaptations of the all-important plant life in this fluid environment. Soon, you will gain a more intimate appreciation of this world through the joy of your own discoveries.

A small blue damselfly and a larger cousin recline on a stem above the surface of a pond.

The Breath Of Life

"Thump-thud!"

The strange sound seemed to come from the solar greenhouse. Now awake, I stepped into the small glassed room, but saw nothing unusual except the glow of a full moon.

"Thump-thud!"

An object bounced on the glass above my head. A glint of light reflected from its oval shape as it rolled down the sloping pane.

Outside lay the motionless body of a large beetle, its hard shell glistening in the moonlight. The insect was a predacious diving beetle, *Dytiscus*. The carnivorous insect had been flying from pond to pond during the night. Normally, moonlight reflected from the surface of the water helped the insect locate another hunting place. Instead, the light reflected from the glazing led the beetle to a collision with my greenhouse.

Dytiscus is a creature crafted for life underwater. Smooth contours grace the hard, mahogany body. Its oval shape slips through the water with remarkable speed and maneuverability and takes to the air with equal strength. But most of its life is spent submerged among the pond weeds waiting to grasp its next victim which it devours underwater. Yet, how can this be? A scuba diver trying to swallow a fish or clam will drown. He cannot swallow and breathe at the same time. *Dytiscus*, however, does not breathe with its face. This leaves the mouth free to eat, and eat, and eat. Even insects living on land do not breathe through a nose or mouth. Instead, a series of holes (spiracles) along the body's side conduct life-sustaining oxygen to the cells through microscopic branching tubes (trachea).

When flying, *Dytiscus* breathes directly through the spiracles like any terrestrial insect. Underwater, however, the beetle must carry a supply of air on its back. Just as the beetle dives, the hard wing cover flicks open and plucks a bubble of air. With the silvery parcel squeezed against the breathing pores, *Dytiscus* can remain submerged more than thirty hours.

The amount of oxygen in the bubble alone is not enough to sustain an actively hunting beetle that long; in fact, *Dytiscus* would drown within minutes if not for an ingenious solution.

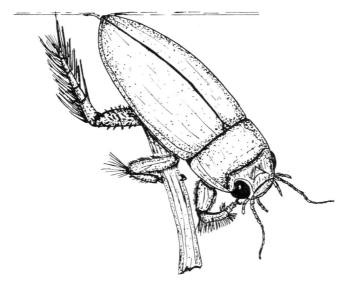

Dytiscus, the diving beetle, clings to a piece of plant stem while waiting for a meal to swim by.

The bubble of air acts like a "physical gill" to draw oxygen from the surrounding water. This replaces the gas used by the insect. Waste gases like carbon dioxide are breathed into the "gill" which pumps them into the pond. Thus, fresh oxygen is supplied to the insect for hours while those gases which will cause suffocation are removed. Eventually, *Dytiscus'* scuba tank shrinks as nitrogen is lost to the water, and he must surface to refresh the bubble.

Many small aquatic insects use physical gills more efficiently than *Dytiscus* and rarely, or never, need to replenish their air bubbles. The aquatic bug, *Aphelocheirus*, is covered with a layer of water repellent hairs. The velvet hairs are closely arranged at the rate of two million per square millimeter. This microscopic fringe traps a film of air so efficiently that the bug lives permanently submerged.

Snorkeling is a method of breathing underwater which man has borrowed from the pond world. Rat-tailed maggots, for example, possess an adjustable siphon which they can extend to four inches. Breathing holes open at the siphon's tip which is protected by water repellent hairs and wax. Thus, the maggot hides its soft body in the mud while a telescoping snorkel exchanges gases. This allows rat-tailed maggots to inhabit polluted waters with low amounts of oxygen.

Other insects also breathe with siphons. The larvae of mosquitoes dangle from the surface film, their tiny siphons drawing in fresh air. The tip of each tube is fringed with water repellent hairs to keep them open. To escape from the holding force of the surface film, mosquito wrigglers give a quick jerk with their swimming brush.

Breathing tubes enable the giant water bugs, water tigers, and water scorpions to remain suspended beneath the surface ready to wrap their powerful forelegs around a careless insect or tadpole. The presence of these ferocious hunters makes pond travel hazardous. Going to the surface for a breath of oxygen can be fatal. Thus, many pond creatures have evolved living gills which allow them to remain buried in the bottom's ooze or hidden under rocks and stumps indefinitely.

Nowhere does Nature's craftmanship show more clearly than in the delicate lacework of gill filaments. As many gill designs

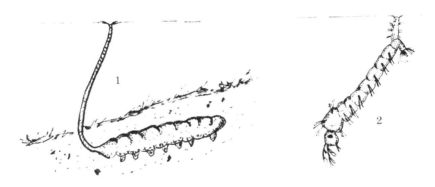

Many insects, such as the rat-tailed maggot (1) and the mosquito larva (2) use snorkles for breathing.

exist as animals to grow them. Some gills fringe the neck, legs, or tails of various creatures. These external gills lie exposed outside the body. Internal gills are buried inside the animal's tissues, protected from the rigors of the environment. In each case, common threads of function and structure tie these gills together. All gills remove oxygen from the water to sustain the cells packed deep within the animal's body. At the same time, waste gases such as carbon dioxide leave the animal to be dissolved in the surrounding water. Frequently, the surface area of the gills is enhanced by thousands of delicate filaments which speed the exchange of large volumes of gas.

Numerous immature insects breathe underwater with external gills. They spend most of their lives as feeding nymphs, developing for months, even years, for a few brief moments as mating adults. Mayflies, for example, exist as adults less than twenty-four hours. So brief is their adulthood that mature mayflies lack jaws for feeding, but they live long enough to mate and lay eggs for future generations.

Mayfly nymphs bear leaf-like gills on their abdomens. The delicate, thin-walled structures are connected to a maze of branching tubes which direct oxygen to the body cells. The slender damselfly nymph carries three feathered gills at the tip of the tail, while dragonfly nymphs may protect their gills in abdominal chambers. Feathery tufts behind the legs of stoneflies diffuse oxygen from the water to sustain the large two-inch nymphs. The wandering caddisfly larvae protect their gills inside a portable home constructed of pebbles or twigs.

Crayfish, too, protect their gills. A hard outer shell (exoskeleton) wraps around the body to enclose twenty pairs of gills. Refreshed water is constantly circulated under the shell by special legs near the head called gill bailers. By watching a crayfish in an aquarium, you can see bits of debris in the water enter beneath the shell and rush out near the head. These particles are being pulled along in currents set up by the gill bailers.

Of all the vertebrate animals living in water, none are more perfectly adapted to their environment than fish. Their scales overlap like shingles to reduce drag and protect the body. Numerous fins provide maneuverability and instant response. But the fish's greatest advantage in the water is the highly efficient gills. Working in unison, the mouth and gill cover (operculum) function much like a pump to bring refreshed water to the gills. With the gill cover tightly closed, the fish creates a partial vacuum and sucks water into its mouth. Oral valves then seal the mouth. If you watch a fish in a tank, you can observe these flaps flick across the mouth. The trapped water is then pushed across the layers of gill filaments which fringe each gill. Tiny blood vessels called capillaries carry their cargo of red blood through the filaments. Oxygen quickly enters the blood and is swept to the two-chambered heart which pumps it to the body. Waste gases diffuse into the water as it flows out of the gill slits.

For many pond animals, life underwater is simply a matter of holding their breath. Turtles fill their lungs through nostrils at the tip of their snouts before diving. No need to surface and be exposed, they merely push their noses from the water, inhale, and submerge. Diving mammals like the water shrew, muskrat, beaver, and otter all hold their breath. The beaver even has tiny valves which seal his nostrils underwater.

Loons, mergansers, grebes, and other diving birds must divide their time in the lake between swimming and diving. For buoyancy, their lungs have many extensions throughout the body. These help them while swimming, but become a problem when diving. To make submerging easier, these birds exhale instead of inhale before going underwater. This reduces their buoyancy. Enough oxygen remains in the lungs and in the extremely efficient blood to allow them many minutes beneath the surface.

If a quick change artist of breathing exists in the pond, the frog earns the name. As a tadpole, external gills draw oxygen from the water, but not for long. Internally, changes are taking place. Lungs are growing and nostrils appear on the head. Soon, the gills begin to shrink, and the frog takes his first breaths at the surface to expand his newly formed lungs. As an adult, the gills disappear, forcing the frog to depend on his lungs. But the frog does have a back-up system. The lining of the mouth and surface of the skin are criss-crossed with millions of microscopic blood vessels capable of extracting oxygen as long as the skin is kept moist.

Amazingly, a few animals breathe with neither lungs nor gills. Simply being moist is adequate. Tiny creatures like the one-celled *Amoebas*, the flatworms, and the *Hydra* gather oxygen directly from the water. Their tiny bodies are thin enough to allow the passage of oxygen through their tissues and into each cell.

The freshwater sponge, an animal composed of millions of amoeba-like cells, constructs a body filled with twisting passages. The cells lie in rows lining the corridors which open to the water through pores. Tiny hairs extending from each cell beat in unison to create water currents through the animal. The currents bring both food and oxygen to every cell.

A beneficial relationship develops between the sponge and a green algae. The sponge provides a home for the floating plant, protecting it from its enemies. In exchange, the bright green chlorophyll of the algae photosynthesizes inside the sponge, and supplies it with oxygen.

The importance of plants in a pond or lake extends beyond a

source of food. The green magic of photosynthesis not only produces necessary sugars, but it also yields the life-supporting oxygen every animal requires. Observe a cluster of pondweeds as they bathe in the sunlight. Minute silvery bubbles of pure oxygen cling to these green factories like strings of pearls. Oxygen dissolves from the bubbles into the surrounding water where gilled animals eagerly breathe it in. Above the water, lungs draw upon the same oxygen. In the aquatic world and the terrestrial environment, dependency among organisms is Nature's way.

Surface Dwellers

You can discover more mysteries and unlock secrets aboard a boat to explore the world of surface dwelling creatures on ponds and lakes. As a youth, I explored this watery world in an old wooden punt.

For years, the punt laid across a pair of sawhorses, her flat bottom collecting hammers and trimming shears, buckets of paint, and baseball gloves. The garage was not the place for a rowboat; she belonged in the water. But, the bottom-turned-workbench had rotted and needed replacing. The rear seat was broken and a gunwale was split. The white paint had yellowed and peeled. Large flecks hung like dried leaves on a dying tree.

My father had wanted to repair the boat but the pressures of his job allowed him little time for such luxuries. However, when a serious illness prevented him from returning to work, he decided repairing the punt would be good therapy for both of us. The following days and weeks were spent removing the rotted bottom one board at a time. Each plank was replaced with a fresh, but identical, piece. The seat was repaired, the gunwale restored. New thole pins were carved from pieces of oak to provide sturdy fulcrums for the oars. Finally she was scraped clean, sanded, and painted white.

The launching was a disappointment for me because our craft "leaked like a sieve." My father was undisturbed and informed me that she would have to "take up a bit." And take up she did. Within a few days we bailed her out and she remained tight and dry.

With a knowing smile, my father asked if I'd like to be responsible for the punt and learn to handle her. My emphatic answer left little doubt in either's mind; thus began my first adventures with a "pleasure craft."

The twelve-foot punt was of ancient vintage and more treacherous than any craft I'd ever been aboard. She would promptly list heavily whenever the uninitiated failed to board amidships. More than once, she confounded my efforts to win her confidence and unceremoniously dunked me. But, with time as my ally and with the persistence of youth, I eventually learned her quirks and shipped the little punt on numerous excursions.

Quiet hours were spent investigating some distant object bobbing on the surface or simply drifting silently with the wind. The only sound was the rhythmic beat of the ripples against the hull. By lying in the bottom of the boat, I could closely approach the spotted sandpipers which scurried along the rocky shore or the dapper mallard drakes as they dabbled bottoms up. Even a sleeping otter was unaware of my drifting approach. In that agile and useful craft, I learned to enjoy the pleasure of peaceful water travel.

Today, my antique wooden punt is retired, but not my desire for unhurried and noiseless boating. To fulfill this need, a far more stable seventeen-foot canoe substitutes for the rowboat and carries our entire family. Not only does the canoe glide soundlessly, but it also adds a new dimension which was lacking with the punt—the canoe is portable. Since it only weighs 75 pounds, the canoe is easily carried on the top of our car and allows us to explore many interesting lakes and hidden ponds throughout New England.

Our most enjoyable paddling is at dusk, after the warm breezes of summer have died. The evening lake mirrors the blues and golds of a fading sunset on its smooth surface. Tall, pointed spruce and fir lose their identity and blend into a jagged silhouette against the sky. The lonesome cry of an unseen loon echoes in the distance. The feeling of solitude is nearly complete until we realize that we are not alone. The calm surface is shared with hundreds, perhaps thousands, of twisting, darting specks of life. Unlike the predictable turns of the canoe, these tiny "speedboats" rush about in dizzying spirals. When an individual speedboat rests for an instant, a shiny black insect takes shape—the whirligig beetle (*Dineutes*). The hard-backed beetle is divided at the water line into two unique halves. The upper half is designed for a life out of water. A carpet of oily hairs repels fluids and helps the whirligig ride high like a bobber. The large compound eyes watch for predators and prey on the surface. When an insect falls on the water, whirligigs respond in numbers to fight over the struggling victim.

The lower half of the beetle lacks the oily hairs but possesses three pairs of legs adapted for swimming. When the whirligig uses his strong muscles to push the legs on the power stroke, flat plates overlap to form an effective oar. As they are drawn forward with the recovery stroke, the blades separate, much like the "feathering" of a canoe paddle, to reduce drag.

Since the whirligig's spinning life is on the surface, it attracts much attention from hungry animals below. To detect these predators before he becomes a meal, the whirligig has another pair of eyes on the submerged half of its body. These eyes focus a second image, separate from the above-water eyes—the insect version of "bifocals."

The whirligig is unusual among surface-dwelling insects because he paddles through the water much like a canoe. Other insects travel on the firm, but flexible, surface. An interesting property of water, dependent upon the attraction of water molecules for each other, allows a dense layer of molecules to accumulate where water and air meet. This layer is the surface film. Not only does the film provide a ceiling to walk, crawl, and hang on, but it also functions like a conveyor belt to move its cargo of life. Winds rushing over a lake or pond create currents at the surface through friction. Tiny passengers clinging to the surface skin can be swept about by this mass transit system.

The water strider (*Gerris*) is one of the numerous commuters on the surface film. Its tapered body is suspended by six remarkable long legs capable of distributing the body weight effectively upon the thin film. The tip of each leg sprouts short oily hairs which are water repellent. Instead of piercing the film, these hairs merely dimple it. Although the strider can literally

Long slender legs support the water strider on a thin film of water.

leap and run upon the water, he usually does not leave his feet. By keeping all six legs in contact with the water, he can row with the very long middle pair, while the other four legs go along for the ride.

When you observe a water strider, watch for tiny concentric rings around the insect. Striders tap out a type of Morse code with their spindly legs to communicate with other members of their species. Messages may be invitations to a possible mate or statements of ownership for a particular portion of the pond.

Ripples on the surface also aid in the location of food. When an insect falls to the water, light reflected by its struggles attracts a hungry strider. If the meat is to his liking, the water strider grasps the prey with his two front legs and drains the body fluids. When the empty carcass is discarded, he balances on a tripod of legs while dirt and dust are cleansed in a ten-minute ritual.

Despite frequent grooming, many striders carry tiny red mites near their head and under their wings. The parasitic mites tap into the body fluids of the water strider and remove nutrients from their host as the host dines on his own meals.

In spite of the strider's mastery of walking on water, the surface film can become a deadly trap. If one is swamped by a wave or in some way is forced beneath the surface, the strider quickly drowns as the surface film forms a barrier which he cannot break through.

One of the largest creatures supported on the surface film is the fishing spider (*Dolomedes*). As large as a half dollar, the brown and black spider is supported on eight long legs covered with water repellent hairs. Without these hairs, the spider would break through the surface skin instead of skating upon it. By rushing onto the film from a dark hiding place, the eight-eyed hunter can surprise and capture surface-dwelling insects.

The hairy legs of the fishing spider allow it to dash across the surface after prey.

Mount Chocorua rises above Chocorua Lake in New Hampshire.

Two large fangs inject a poison into the struggling victim to paralyze it. Unlike the strider, the fishing spider is not restricted to the surface film. By grasping a floating leaf, the spider can pull itself under water to hunt fish and tadpoles. A silvery coating of bubbles trapped in its hairs allows the spider to remain submerged for nearly an hour.

Early morning paddling on a glassy lake also is a unique pleasure which can awaken even the most tired senses. Clouds of vapor wisp above the dark water and veil the sun. Phantom outlines of evergreens mark the boundary of land and lake. Sounds travel freely and clearly in the dense morning air, making their locations difficult to pinpoint.

As the sun burns through the mist, the yellow heat warms the back of your neck. A few frantic frogs plop into the water, only to surface beside a tangle of plants. A yellow warbler sings from a clump of alders; the lake whispers past the gliding canoe. Ahead, at the edge of the swirling wisps, a low, dark form emits an eerie call, then disappears beneath the surface. The high-pitched tremolo of the common loon (*Gavia immer*) echoes across the lake. His warning cry is answered by a distant loon; then, all is silent.

Deep in the mist, a rhythmic splashing tells a story of escape. Out of sight, the loon becomes airborne, but not without considerable effort. Loons have a small wing area compared to their

A common loon, displaying its distinctive "necklace" of feathers, warms the eggs on its island nest.

large, heavy bodies. As a result, they must run across the surface of the lake while flapping their wings. Sprints of a hundred yards are not uncommon. The splashing dash is awkward, but once in the air the loon flies powerfully and rapidly, reaching speeds of 60 miles per hour. Loon landings are even less graceful than their take-offs. Since the small wings cannot slow them enough to make a soft landing, the bird simply skims close to the surface and belly-flops to a splashing halt.

Walking is even more of a struggle because the legs are located far to the rear and are encased by the body to the ankle. Hence, a common loon maneuvers with a clumsy waddle while trying to support the front of the body with bill and wings.

It would seem that a loon is a poorly designed creature until the bird is placed in its preferred environment. Once in the water, you'll see how liabilities of air and land become assets for aquatic life. Powerful leg muscles and webbed toes blending perfectly with their aft position propel the loon smoothly and rapidly. The solid bones, a hindrance to flight, provide ballast for submerged hunting. Extensions of the lungs throughout the body compensate for the bones' weight and buoy the loon while swimming at the surface. By exhaling this extra air, a loon can sink quietly. By arching its slender neck, it can dive quickly.

Beneath the surface, loons pursue fish of all kinds plus frogs and leeches. Dives of about thirty seconds are common, but longer dives of three minutes are possible. Submerged times of ten to fifteen minutes are probably accompanied by brief trips to the surface where the loon snorkels with only the tip of his bill exposed.

The breeding plumage marks the beginning of another mating season for loons which remain paired for life. Returning to fresh water as soon as the ice is out, the pair establishes a territory of about one hundred acres away from the influence of people (hopefully). Their need for privacy is demonstrated by dramatic displays of raised plumage, billdipping, diving, and up-right positions to intimidate intruders. Although these methods work well on other loons, people find them fascinating and inadvertently cause serious problems if the loons leave the nest unprotected. With continued trespassing, loons may abandon the nest completely.

Where privacy is respected, loons lay two dark eggs during the spring. The nest is located near the water to reduce walking and to provide a quick escape route. A small island, hummocks, floating mats of plant debris, or even a muskrat house can serve as a nesting site. Once established, the same site will be used each year.

Both parents take turns brooding the eggs for thirty days until the dark brown chicks hatch. The downy hatchlings take immediately to the water, ready to swim. The parents teach the young to dive and to capture fish. When the chicks tire, they hitch a ride on the parent's back. By September, offspring learn to fly and the family breaks up.

During the fall, the adult breeding plumage is replaced by drab gray and white: the winter colors. This plumage will last through the winter which loons spend along the ice-free coast in silence. With the return of warm weather, however, they return to fresh water to haunt the mists of early morning with their mysterious cry.

Loons are not the only birds to dive in lakes and ponds. Grebes and mergansers possess many similar adaptations and spend much time hunting beneath the surface. Mergansers even have saw-like teeth on the edges of their narrow bill to aid in the capture of fish.

An array of colorful ducks feed and breed throughout the freshwater world. Emerald-headed mallards, golden-eyed whistlers, fast flying green-winged teal, and the blue-billed ring

Hunted to near extinction during the beginning of this century, the wood duck has made a strong comeback.

neck are but a few of the dabblers and divers dependent upon our wetlands. Handsome as these ducks may be, their plumage wanes compared to the distinctive wood duck drake (*Aix sponsa*). The wood duck drake's crest glows with iridescent greens and purples. Crimson eyelids and irises are surrounded by bronze and burgundy. White pin stripes line the crest. The multicolored bill is tipped with black, edged with white and yellow, and based in red. The body is covered by a profusion of tiny black lines, bold white stripes, and shimmering bronze, purple, and green. These appear painted by an imaginative artist experimenting with his pigments.

By comparison, the female duck is drab indeed, her colors muted browns and greens, with only a hint of her mate's iridescence. But, the hen has an advantage in a loud squealing voice which summons her mate whenever she leaves the nest.

Wood ducks select suitable cavities in oak, elm, and maple trees for nesting sites. The final decision is made by the female who inspects numerous holes, rejecting shallow cavities and narrow entrances, until the appropriate site is found.

As the dull white eggs are laid in the nest, the female plucks down feathers from her breast. When the clutch is complete (as many as 15 eggs), enough down is added to cover the eggs during the hen's "nest-break." This down comforter prevents the eggs from being chilled. The clutch is incubated for about 30 days. The dark-backed hatchlings are brooded for a day before leaving the nest.

The day following hatching, the female carefully inspects the area around her nest for danger. When she is convinced that all is secure, she flies to the ground and calls to her babies. In response, the young ducklings jump to the entrance hole, pause, and then leap to the ground. Despite a 10 to 30 foot fall and tumbling landings, the ducklings survive and follow their mother to the nearest lake.

Frequently, the nesting site a wood duck chooses is inadvertently provided by North America's largest rodent—the beaver (*Castor canadensis*). By constructing dams and flooding several acres of woodland, beavers drown scores of trees. The more flavorful trees are gnawed down for food and building, but others remain standing as rotting snags. Woodpeckers in pursuit of a varied insect menu, riddle the snags and also carve out nesting cavities. These abandoned chambers, when large enough, are utilized by wood ducks.

Tree swallows also rely on the excavations of the woodpecker for nesting pockets. These graceful fliers time their spring return to coincide with the first hatchings of insects. Twisting and gliding on iridescent wings of green and blue, they snatch insects in flight and carry them back to their nests. With acrobatic precision, swallows skim the pond dimpling the surface with their beaks to drink or to capture a tiny bug on the water.

Unaware of their benefits to these birds, beavers erect their dams for only one purpose—self-preservation. The impounded water protects the beavers from predators and provides a storage area for winter supplies of food. One enemy, however, is little deterred by the flooded acres and actually utilizes the water to help trap the big rodents. Highly prized for their valuable fur, beavers are trapped today by man as they have been since America's colonization. However, 80 years ago beavers were nearly exterminated from most of the northeast. Centuries of relentless trapping in search of the "brown gold" had diminished an estimated population of 60 million animals to near extinction. Finally, Congress enacted legislation in the early 1900s to give protective status to the exploited mammal which has since made a remarkable recovery.

Part of the beaver's dramatic comeback is due to his resourcefulness and part to his tendency to overpopulate. Mating in late winter, a litter of three to six kits is born in April or May. These young beavers are raised with parental guidance for nearly two years. Then, the parental bond dissolves and the youngsters are chased from the lodge to establish their own colonies. A typical beaver colony may have fifteen members including the two parents, a litter of kits, and immature animals about a year old.

Beaver leave numerous signs of their presence, the most obvious being their dam. The confusing mass of branches and mud is actually constructed carefully and instinctively to yield a strong, durable barrier.

Working as a team, the beavers topple aspen and popple trees in quantity. The sweet bark is peeled for food and cuttings are dragged to the dam site. On a foundation of rocks and mud, layer upon layer of branches and tree trunks are crisscrossed, their butt ends pointing upstream. Mud, rocks, and soggy plants are dumped into the cracks to cement the dam together.

The beaver pond quickly swells behind the barrier, usually flooding about eight acres to a depth of four feet. Attention is then turned to the construction of a suitable lodge, using the same building materials and techniques as the dam. The jumbled hump of sticks rises several feet above the water. All entrances are built below the surface making winter exits beneath the ice possible. Inside, a living chamber provides a dry bedding area for sleeping and for raising the kits.

Like all engineers, the beaver reports to work with the proper equipment to do the job. A thick coat of glossy brown fur keeps it warm. By oiling the fur regularly, it remains water repellent. The oil is spread during grooming by a "split claw" on the hind foot. The hind feet are webbed for efficient swimming, and the flat scaly tail propels and steers the chunky body. When the tail is slapped hard against the water, the loud smack warns other beavers of approaching danger. If necessary, a beaver may remain submerged for nearly a quarter of an hour and swim one-half mile under water.

Probably a beaver's most important tools are his dexterous forefeet and chisel-like incisors. The long front claws can dig in the mud and manipulate the whittled cuttings. A strong little finger is adapted for gripping and compensates for a weak thumb. The beaver's teeth are both a necessity and a threat to survival. Although the two pairs of orange-red incisors are indispensable for chopping trees, peeling bark, and gripping cuttings, their continuous growth must be held in check by constant

A beaver lodge, composed of mud and sticks, protects the builder from his enemies and the weather.

use. If they grow too long, the teeth can prevent feeding, and starvation follows.

Once settled, the influence of a beaver colony on its surrounding environment and wildlife is surpassed only by the activities of man. The flooded acres attract teal, ring necks, mallards, and wood ducks which feed and dabble on the flowage. Salamanders and frogs breed in the shallows, and toads leave long strings of eggs draped across the numerous water weeds. Spatterdock sprouts near the shore. Wading moose graze on the spatterdock which they share with the pond's creator. Cattails and sedges attract the beaver's aquatic cousin, the muskrat (*Ondatra zibethicus*). This web-footed rodent constructs a home of cattails, but it is more modest than the beaver's elaborate lodge.

Insects quickly populate beaver ponds. Many like *Dytiscus*, locate the flowage by reflected moonlight and "drop in" to search for meat. Fish, particularly brook trout, inhabit the waters as long as the temperatures remain cold. When the flowage warms, the trout move upstream to be replaced by more tolerant fish.

Broad, S-shaped ripples on the surface betray the presence of the common water snake (*Natrix sipedon*). The stocky reptile hunts fish, crayfish, and frogs, and seizes the prey with sharp, non-venomous teeth. Easily angered and quick to defend itself, the water snake should be left alone. Contrary to popular belief,

The northern water snake is a beneficial, non-venomous reptile.

they do not destroy valuable game fish or compete with fishermen. Instead, the snake catches slow-swimming chubs and minnows. Any game fish they catch are usually injured or diseased, and their removal benefits the healthy fish. Food is held securely by the snake with its backward pointing teeth. By stretching the ligaments connecting the jaws, sizeable prey can be swallowed whole. Once in the stomach, powerful acids dissolve the meal.

Being a reptile, the water snake must bear its young on shore. As many as 24 to 48 squirming young are born alive in late summer. Like the adults, they flick their forked tongues into the air to "smell" their environment and to seek clues to danger and to food.

Occasionally, the tables are turned on these smaller snakes by the voracious bullfrogs (*Rana catesbeiana*). The large frogs attempt to swallow anything that moves and cannot swallow *them*. Dragonflies, small snakes, other frogs, flies, minnows, even tiny birds and mammals are snared by a muscular tongue flipped from the cavernous mouth.

Eventually, the beavers exhaust the reserves of popple and aspen and seek fresh supplies elsewhere. The abandoned sanctuary lingers behind the durable dam for many years. Without the channeling efforts of the beavers, however, the pond gradually fills with silt. Dying, as all ponds must, the flowage changes into a beaver meadow. Grasses, berries, sedges, and shrubs sprout from the rich stream deposits behind the decaying dam.

Where dainage is poor, the decay of accumlating plants is slowed. Acids build up and sphagnum mosses invade the dying pond. Layers of peat may collect in the old basin, turning the flowage into a bog.

Finally, pioneering trees like the spruce succeed the grasses and mosses. Nourished by the loamy sediments, a young forest sprouts skyward. Someday, perhaps, a pair of beavers may wander into the forested meadow to construct a dam, completing a natural cycle of birth, death, and rebirth.

Bottom Dwellers

Clutching a large glass jar tightly, my son scurried along the weedy shore of one small lake. Water sloshed inside the container. Drying fragments of algae clung to the lid. Breathless and excited, he hurried to our campsite and rushed into our tent.

"Dad, come quick!" he blurted. "Look what I've caught! It's a little lobster!"

As the jostled bits of mud and leaves settled, the shape of a strange dark creature began to appear. Two long tappered antennae swept the water for chemical clues. At the base of these segmented sense organs, a pair of black eyes periscoped through the debris. Protruding from the murk, two oversized pinchers drooped in front of the pointed head.

Crayfish, also known as crawdads or crawfish, resemble their marine cousin, the lobster. Growing from three to six inches in length, crayfish are protected by a hard outer covering (exoskeleton). This jointed armor allows the wearer limited freedom of movement and causes definite problems for the growing crustacean. Trapped inside a non-stretch suit, crayfish must periodically shed the crusty skin and grow more accommodating outfits. This molting process usually happens during the spring and fall. Following shedding, the "soft shell" crayfish, named for their new suits which have yet to harden, hide under rocks and submerged debris to avoid their numerous enemies. Stripped of their most important defense, crayfish can easily be devoured by hungry bass and raccoons. Additionally, the strenuous shedding saps much of their strength. Pinchers dangle helplessly, and rapid escape, normally supplied by a flip of their powerful tails, is impossible until the new shell hardens.

Hiding is a natural part of crayfish behavior. Even when they are not molting, crayfish burrow into the bank or crawl under rocks to avoid sunlight and predators, including man.

Under the cover of darkness, crayfish actively seek their food. Mainly scavengers, these freshwater crustaceans clean up dead fish and plant debris from the bottoms of lakes and ponds. They also eat insect larvae and snails. Their two powerful pinchers capture and shred the food. Scurrying on eight legs gives the

With long antennae and dangling pinchers, a crayfish searches for a dead fish.

crayfish a clumsy appearance; yet, their highly coordinated muscles move one leg after the other in a wave of motion sweeping from front to back. Beneath the jointed abdomen, small fringed swimmerets help steer the animal. In May and June, the swimmerets become a cradle for nearly 800 eggs. Protected beneath her armored tail, females "in berry" keep the eggs clean and supply fresh oxygenated water. When hatched, the miniature crayfish remain with their parent for a short time before starting life on their own. Although the crayfish makes an excellent parent, she really has no choice in the matter since her eggs adhere to her swimmerets.

Another bottom dweller, the brown bullhead (*Ictalurus nebulosus*) diligently cares for her young though she could easily abandon them to their fate. In late spring, bullheads gather to spawn in the shallow water at a lake's edge. The "whiskered" male selects a nesting site of gravel or sand and sweeps it clean. His mate drops bunches of sticky eggs into the cleared nest, each containing about 2,000 eggs. Under favorable water conditions, the cream-colored eggs hatch in less than one week to surround the devoted parents with a cloud of tiny black fish. Both parents vigorously protect the newly hatched fry against all intruders, even to the extent of engulfing the babies into their mouths until the danger has past. Like a shepherd tending his flock, the parents herd the young fish together until they grow to nearly two inches.

Eventually the instinct of guardianship wanes and the adult hornpouts abandon their young. The fry live out the remainder of their first year hiding among the plants near shore.

When fully grown, these fish may reach eighteen inches. The deeply pigmented skin lacks the crust of scales found on most fish. But the smooth, soft body and fins belie a hidden weapon. Concealed behind the gills in a pair of innocent looking fins are two stout spines capable of delivering a nasty prick to the unwary. A third spine hides in the dorsal fin atop the back. All three spines have venom glands which cause a painful, but not dangerous, wound. The real threat lies in possible infection from bacteria associated with the "horns."

Surrounding the mouth, a series of whiskers gives the hornpout his cat-like countenance, and aid the bullhead during his nocturnal search for food. Sensory nerve endings scattered through the barbels locate a variety of food items by touch. Plants, insects, crustaceans, fish eggs, and small fish are all eaten. In fact, their gluttony makes the bullhead an easy fish to catch. At night, a large wiggling night crawler left on the bottom invariably results in a tug on the line and a fish in the boat.

When thinking about bottom-dwelling pond creatures, catfish and crayfish come quickly to mind. They are easy to catch and both taste good. Yet, these two bottom-dwellers are vastly outnumbered by a veritable zoo in every square foot of bottom muck.

Collecting other tinier sediment dwellers is not difficult. With the aid of a simple hand lens, tweezers, a white enamel tray, and collecting dredge, many of these animals can be captured and observed. Simply rake the dredge through the bottom ooze near the shore and drag up a small amount of the rich debris. Half fill the white enamel pan with clean water and add some of the sample. Sort through the material with a pair of tweezers, using the hand lense when necessary. Many of the intriguing animals you find can be identified with a freshwater field book.

Tiny worms in a variety of shapes abound in the pond's ooze. Nematodes, or roundworms, probing the mud with their thread-like bodies, feed on tiny plants, bacteria, and protozoans. When swimming, nematodes are recognized by the S-shaped kinks of their slender bodies. Sharing the bottom are numerous segmented worms. Some, like *Aeolosoma*, are nearly transparent and dotted with tufts of bristles. These live and feed on bottom

52

debris. A cousin, the large-mouthed *Chaetogaster*, devours insect larvae, tiny crustaceans, and smaller worms as it burrows through the sediments. *Tubiflex* worms spend their lives standing on their heads which are buried in the mud. Their burrows extend above the bottom as fragile tubes. Waving their reddish brown tails helps to circulate fresh oxygenated water into the burrow.

Hidden in the jungle of roots and decaying leaves lurks an interesting, although unsavory, character—the leech. The segmented leeches are generally parasitic on fish, turtles, snails, frogs, and aquatic mammals. Although the common name, bloodsucker, describes the habits of most leeches, others live freely.

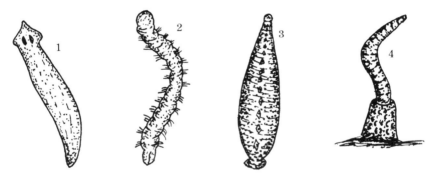

Some common worms found on bottom sediments are (1) *Planaria*, (2) *Aeolosoma*, (3) *Macrobdella*, (4) *Tubiflex*.

These flattened worms frequently attach to rocks and sticks on the bottom and stretch their bodies outward to search for a victim. Two suckers, one at each end, permit the leech to remain attached at the tail while the head and body wave in the water like a length of ribbon. When a frog or fish swims near, the mouth sucker quickly attaches to the prey. A vacuum action secures the mouth sucker, and sharp teeth cut through the skin. A hollow tube pierces the wound and draws the blood into the leech's chain of stomach sacs. A special chemical found in the leech's saliva keeps the blood from clotting and helps preserve the meal for as long as eighteen months.

A common bloodsucker, *Macrobdella*, grows fairly large and swims gracefully. By stretching the elastic body, an undulating swimming motion can carry the leech quickly toward a victim.

A chain of orange splotches on an olive green back help identify this parasite.

The pond's sanitation department resides in the silty residue of bottom sediments. Here, microscopic bacteria compost the constantly accumulating mounds of organic matter. Aiding the one-celled bacteria are armies of scavenging animals. One, the arrow-shaped *Planaria*, "tastes" the water with its chemical-sensitive body. The abundant flatworms, less than an inch in length, glide across the sticks, rocks, and silt on a carpet of coordinated hairs (cilia) searching for a dead worm or a dying tadpole.

Among the miniature inhabitants which live in the ooze and dine on the all-purpose bacteria resides the *Amoeba*, a "walking puddle." To the unaided eye, this drop of life resembles a speck of dust, but under the microscope a marvel of organization appears. Housed within a transparent film, a complete individual breathes, eats, moves, replicates, and dies. Movement is accomplished by flowing currents of fluid inside its flexible covering. These currents stream forward into "false feet" which stretch ahead drawing the hind end with them.

While the *Amoeba* plods along in slow motion, the *Paramecium* churns ahead propelled by miniscule oars called cilia. Bouncing off immovable objects, the *Paramecium* blunders forward in a spiralling motion. As it zigs and zags throughout a drop of water, the "slipper-shaped" *Paramecium* catches bacteria and plants in its cilia-lined mouth.

Paramecium and other one-celled organisms become food for a curious animal, the freshwater sponge (*Spongilla*). Frequently thought of as a plant, the sponge is a confusing collection of cells arranged into a series of tubes and openings. Supported by stiff

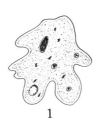

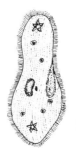

Two microscopic pond dwellers are the slow-moving (1) *Amoeba*, and the speedy, but erratic, *Paramecium (2)*.

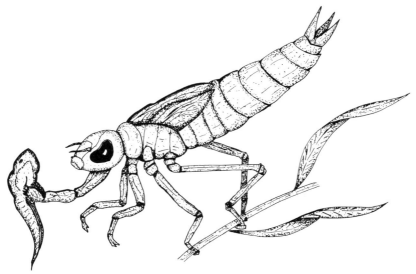

A voracious dragonfly nymph captures a tadpole with its unusual lower jaw.

rods, each tube is lined with cells waving a whip-liked organelle, the flagellum. The flagella beat against the water creating currents which sweep food and oxygen through the twisting passages.

Sponges are generally drab brown or tan, but frequently bright green sponges can be found attached to submerged sticks. Green algae (*Chlorella*) living in the sponge colony causes the color change because of its chlorophyll. This beneficial relationship provides a secure home for the plant which shares its food and oxygen with the sponge.

Insects and tiny worms hide inside the passageways between the finger-like growths of sponges. They receive the benefit of concealment and convenient feeding as the sponge circulates water and plankton through its body. The numerous tenants found associated with the sponge also attract various predators. The bulging-eyed nymphs of the dragonfly regularly prowl the surface of sponges, rocks, dead leaves, and logs in a constant search for food. The torpedo-shaped nymphs catch any animals they can handle: fish, worms, tadpoles, and other insects.

If you dredge up a dragonfly nymph, place it in a container of pond water for observation. By feeding it small worms, you can watch how the predator captures his prey. The large compound eyes project a mosaic of images into the nymph's brain which is particularly receptive to movement. As the worm wriggles about, the nymph approaches stealthily until it reaches striking distance. In an instant, the lower lip which has been neatly folded beneath the head, flashes forward and seizes the prey with hook-like teeth. Securely impaled, the victim is drawn beneath the mouth and devoured.

When speed is required, the nymph tucks its legs against the body and zips about by "jet propulsion." It simply draws water into the digestive system through the terminal end. Muscles squeeze the water with great pressure. As the water is suddenly expelled, the insect spurts forward.

When the nymph has matured, it climbs from the water on the sticks. At the moment of transformation, the hard skin splits down the back and the adult dragonfly struggles free. In a short time, the abdomen expands and straightens. The miniature wings quiver and stretch from the blood coursing through their veins. Finally, the adult dries its wings and body and takes flight. Although the adult is freed from gill breathing and hunting beneath the water, one link still ties the dragonfly to the

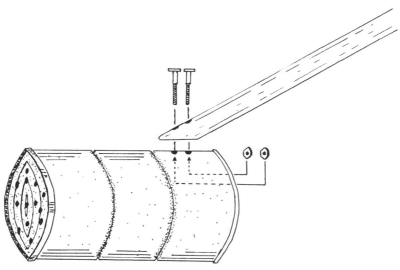

A simple, but effective dredge can be constructed from a number 8 can and an old broom handle. First poke several holes in the bottom of the can to let excess water drain out. Next, drill two holes in the handle for a pair of bolts and attach as shown in the diagram. To use your dredge, drag it slowly towards you allowing it to scoop up leaves and other bottom material, including many small creatures.

fluid environment—egglaying. After the mating flight, the female dragonfly hovers over the shallows and dips her abdomen in the water. Some species simply drop their eggs to the bottom while others conceal them among floating leaves or on plant stems.

As the ugly dragonfly nymph gives little evidence of the marvelous change into graceful adulthood, the life history of the freshwater mussel is equally varied. Pearly mussels bear as many as three million embryoes on their gills at one time. Each embryo, called a glochidium, only slightly resembles its parent. Like the adult, the tiny mussel has two shells (valves), but each shell is armed with sharp spines and a twisted thread. When they are ready, small bunches of glochidia are spewed into the water through the parent's siphon. Some are swept away by currents; others settle on the silt. Glochidia are extremely sensitive to touch and are very light in weight. Whenever a fish swims near the waiting larvae, the swishing tail sweeps the glochidia into the water. They immediately begin snapping their toothed shells together. If by chance they clamp onto the fish's gills or tail, the glochidia remain locked in place to become tiny parasites. The tissues of the fish soon cover each larva with a layer of cells to protect them during development. While stealing food from the cells of their host, the glochidia transform into miniature mussels. Within a month, the newly developed mussels drop from the fish and take up life in the sediments.

Adult pearly mussels are living strainers of a freshwater community. Daily, enormous amounts of water are drawn through the incurrent siphon, filtered by the gills, and expelled by the outcurrent siphon. Rows of cilia collect microscopic bits of food and plankton from a sticky layer of mucus on the gills and push the food to the mouth.

You can easily observe this siphoning action by collecting a mussel and keeping it in an aquarium. Once the animal adjusts to its new location, the shells part and a pair of siphons poke into the water. One siphon has a smooth edge while the second is fringed. Add a few drops of food coloring to the water near the fringed siphon (incurrent) and watch the action of the currents which carry food and oxygen inside the mussel.

A bottom-dwelling insect, the caddisfly, carries his house wherever he wanders. Hatched from a string of submerged eggs, the caddisfly larva constructs his custom-built home by first weaving a foundation of silky threads. Once the tube is spun, tiny twigs, sand grains, pebbles, pieces of shell, or bits of

Caddisfly larvae crawl about the bottom in homes constructed from available material such as twigs or small pebbles.

plant are collected and diligently attached to the foundation. Great care is taken to fit the pieces precisely so that the camouflage is complete. If the caddisfly fails in his deception, a hungry trout will enjoy a tasty meal.

When the long days of early summer arrive, the caddisflies hatch from their concealed homes, crawl up emergent stems, and take wing as moth-like adults.

Probably no pond creature has a worse reputation than the snapping turtle (*Chelydra serpentina*). The large, muscular reptile reputedly chomps through broomsticks, devours large numbers of game fish, and grows to a hundred pounds. The snapper is ugly enough to make these fears seem true. A moss-covered shell, warty skin, powerful clawed limbs, musky odor,

Although not as formitable as a snapping turtle, this musk turtle has a sharp beak.

hooked beak, and the presence of numerous leeches dangling about the head do not conjure up pleasant visions of the common snapper. Yet, despite his appearance, most snapper legends are more fantasy than fact.

Although the snapper does have a sharp beak, broomsticks survive the turtle's chomp with little more than a dent. This does not imply that a snapper's mouth is a good place for fingers and toes because a snapper out of water does lunge with surprising quickness. In the water, however, the turtle's only thought seems to be escape.

Before snappers became a heavily hunted gourmet delicacy, turtles weighing forty to fifty pounds were not uncommon. Today, a thirty pounder is unusual. A larger southern relative, the alligator snapper, may occasionally reach two hundred pounds and four and one-half feet in length.

Snapping turtles eat a wide variety of foods. Snails, crayfish, leeches, frogs, fish, and plants are all part of the menu. Lying motionless on the bottom, the serrated shell of a snapper resembles a ragged stone. The mossy growths of algae further camouflage the predator. When a bottom-grazing sucker drifts too close to the patient hunter, the horny beak lashes out with surprising quickness to seize the careless fish. Since fish comprise a large portion of the diet, fishermen often become irritated, but needlessly so. Turtles generally eat "trash" fish and provide a free clean-up of these animals which compete with the important game fish. Research has also shown the number of young ducks taken by predatory turtles is not significant.

A large bullfrog waits patiently for a meal.

Life Betwixt And Between

There's a wild world to explore between the surface and lake bottom, too. You can discover more secrets on a snorkel trip beneath a lake. Here's what I discovered on *one* underwater snorkel dive.

Shafts of greenish-yellow light flicker among the leaves of pondweeds below the surface. Fractured by the rippling surface, sunlight paints a shifting pattern on the sandy bottom. Small schools of fallfish (*Semotilus corporalis*) drift among the weeds. A single black band marks each of the sleek minnows. These young fish scull with pectoral fins and confront explorers with wary curiosity. In graceful unison, a school darts left or right, up and down, forward and even backward to keep a safe distance between us. A predatory pickerel intent on a meal of minnow will have difficulty sorting out a victim from this school of fleeing morsels. Certainly, the old adage "safety in numbers" applies to the survival of these young fish.

Submerged near the base of the pondweeds rises an unusual mound of stones. A closer inspection reveals an irregular pyramid about two feet tall and twice as wide. The neatly piled rocks, some as large as plums, are the remnants of a fallfish nest. An energetic male had constructed the stony cradle during the spring as part of the mating ritual. Perhaps the young chubs hiding in the weeds above hatched from eggs deposited among these rocks.

Ahead, a tangle of flexible stems stretches from the bottom to a canopy of lily pads. Shadows darken the soft bottom. A flash of silver appears near a lily's root stalk and vanishes. Again the sliver of light flickers, reflected from a three-inch fish carrying a piece of pondweed in his mouth. The stickleback darts toward a cluttered pile of plants and sand to deposit the leaf. After disappearing among the weeds, the brightly marked male returns with more material. His silver back and blushing abdomen display the male stickleback's breeding colors.

This particular male is occupied with an urgent task: construction of a suitable nest. Beneath the mound of stems and leaves, he has already dug a shallow pit and lined it with bits of plants. When he completes the roof, the male will escort an interested female to his nest. Inside, she will lay her eggs as the male pokes her side with his nose. When the eggs are safely deposited in the nest, the male will chase her away, fertilize the eggs, and take the responsibility of protecting them from all intruders. The dutiful stickleback will even extend his parental care to the fry until they can live by themselves.

Unlike most fish, the stickleback's body is armored with thin bony plates instead of overlapping scales. A row of stout spines lines the slender back. Depending on the species, these number as many as nine, to as few as two.

Gliding from the shadows of the water lilies, the powerful body of a smallmouth bass (*Micropterus dolomieui*) sweeps beneath me with an air of confidence. The gray-green figure is fully as long as my flippers. He samples bits of debris stirred from the bottom by my artificial fins. Quickly recognizing the delicacies, the bass swallows each with unhurried dignity. However, when a pair of slender antennae and beady black eyes poke from the swirl of silt, the bass loses all composure. He darts headlong into the soft bottom to inhale the hapless crayfish, silt and all. With all dignity lost, the bass swims away as clouds of dirt stream from his gills, but a crayfish fills his stomach.

The smallmouth quickly fades from view despite my best efforts to follow him. During 400 million years of evolution which

A young smallmouth bass cruises near the shore.

A male stickleback coaxes his mate to lay her eggs in a nest he has constructed.

have perfectly adapted fish to their watery world, many body shapes have evolved including the most common and nearly ideal torpedo-shape. This streamlined form slips easily through the water. It is covered with a layer of shingle-like scales and a slimy coating of mucus. The mucus is produced by special glands in the skin and helps protect the fish from bacterial infection. If you catch a fish and intend to release it, wet your hands before touching it. Dry hands remove large quantities of the mucus and make the fish more susceptible to disease.

The gills filter oxygen from water coursing past their filaments with remarkable efficiency. When the oxygen is combined with digested food, energy is released to fuel the overlapping muscles which move the fish through the water so effortlessly. Waves of muscular contraction flowing backward twist the body from side to side. As the body pushes against the fluid environment, the fish is propelled forward, much like an ice skater pushing sideways with alternating thrusts of his legs.

Maneuverability is controlled by numerous fins scattered about the body. Soft rays or stout spins tucked into these folds of skin help stabilize, steer, and power the swimming fish. By sculling a pair of pectoral fins behind the gills, a fish can glide forward or backward.

Chain pickerel frequently approach their prey moving only these pectoral fins. Materializing like a phantom from beneath the shadows of the bank, a slender pickerel steals close to his victim. Hardly a vibration stirs the water as a barely perceptible motion carries him forward. Finally, with its peg-toothed jaws only inches from an unwary minnow or frog, the predator lunges forward to seize his meal. An experienced pickerel has learned the value of patience in hunting for a hasty approach results in an empty stomach.

All fish, whether victim or victor, are extremely sensitive to their surroundings. Although a visible ear is lacking, fish can all hear well. Sound waves traveling through the water can pass through a fish's body to stimulate the inner ear embedded in the bony skull. Additional receptors are housed in a fine streak which lines the fish's flanks. Beneath a series of pores in this streak lies a nerve-lined canal. This structure, called a lateral line, is capable of detecting sounds including the low frequency tail sweeps of potential prey or predators. The lateral line even provides enough information for a fish to swim blindly without bumping into objects.

Fish eyes are constructed much like our own, having a lens, pupil, iris, and light-sensitive retina; however, they lack eyelids. Fish can never blink, nor do they need to. Blinking cleanses our eyes of dust and moistens them against damaging heat and light. The fish's environment does this for them.

The large round eyes float in bony sockets and perceive separate color images. While one eye focuses on a meal ahead, the other eye can look behind or even sideways. Predatory trout and salmon can also focus on a distant spot with both eyes, to give depth of vision and aid in hunting.

A pair of cup-shaped holes on the end of the snout function as the nose. Lined with nerve endings, these blind sacs collect and sort out the wide variety of chemical messages in the water.

The sense of smell plays an important role in a fish's life. Many locate their food by smell, especially in murky water. Minnows, when injured, release an alarm odor easily detected by other members of the school. The warning scent quickly scatters all the minnows and sends them into hiding. Guided by a chemical memory, salmon return to the location of their birth in order to spawn.

There's more to discover betwixt and between in a lake. Try sweeping a fine mesh net through the submerged weeds near a dock or shoreline. You'll find various curious creatures. One typical insect might be a water boatman that strokes through the water with long, flattened legs resembling the oars of a

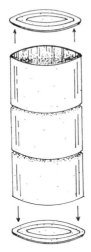 

Spying on life beneath the surface does not always require a face mask and snorkel. By constructing a simple underwater viewer, you can wade in shallow water and watch what is happening below.

Cut the ends from a #8 juice can and tape any sharp edges. Cover the bottom and sides with a heavy, clear plastic wrap. Kitchen wrap will do the job, but if you can find a heavier plastic, it will work better. Secure the wrap near the top of the can with tape or a strong elastic.

Your underwater viewer is now ready. When you place it in the water, tip the viewer to one side to prevent air bubbles from being trapped. Don't be surprised if something is looking back at you!

rowing shell. The grayish-brown body is outlined by a glistening drop of air which serves as its aqualung. In the pond, the boatman uses its small, fringed forelegs to strain bits of plants from the muddy bottom. The second pair of legs clings to weeds and sticks to keep the buoyant insect beneath the surface. The last pair of legs, adapted to swimming, propel the water boatman away from numerous enemies.

A larger relative, the backswimmer, also rows with special oar-like legs. But as the name implies, they travel on a deeply keeled back. These predatory insects occasionally attack small fish but generally dine on other insects. Their mouth is designed to pierce the body of their prey and drain its juices. They can inflict a painful bite with their pointed beak to anyone who handles them carelessly.

Both the water boatman and backswimmer are powerful fliers and travel to different ponds during the night. This useful

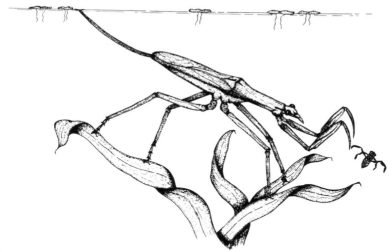

A stealthy hunter, the water scorpion prepares to seize an unfortunate water boatman.

adaptation allows them to escape from small ponds which become too warm or dry up.

You may also find a spindly-legged water scorpion hidden among some leaves. His dark brown body, slender as a match stick, blends deceptively with the background. Moving deliberately, the pop-eyed insect elevates the twin filaments at the tip of his abdomen. As they pierce the surface, interlocking bristles join the two strands into a snorkel.

This pose is also assumed by water scorpions hunting along a lake shore. Clinging to weeds and breathing through the snorkel, scorpions may wait patiently for hours until a water boatman or mosquito larva swims near. Forelegs held in a "praying" position flash out and embrace the unlucky victim between razor-edged legs.

Equally as long, but much bulkier, the giant water bug hunts in a similiar way. More powerful than his skinny cousin, the water scorpion, the giant water bug's menu includes insects, tadpoles, frogs, and even fish.

Voracious as the giant water bug is, he is no match for the greedy water tiger. The water tiger is the offspring of a diving beetle. Like his parents, he enjoys eating. The water tiger swims with powerful strokes of his six bristly legs. The streamlined body is joined to a flattened head bearing two sickle-shaped jaws. These sharp, hollow fangs seize and puncture tadpoles, fish, and other diving beetles. Powerful juices flow through the curved hypodermics into the body of the victim to liquefy the organs and muscles. The *Dytiscus* larva then sucks the fluid back into its own stomach, discards the empty carcass, and seeks a fresh meal.

Mosquito larvae are common in ponds. Hatched from rafts of floating eggs, the segmented wrigglers chew on bits of algae or diatoms and change plant tissue to animal muscle. When eaten by minnows and perch, the mosquito protein becomes fish protein.

Breathing with a snorkel, mosquito wrigglers hang quietly from the surface film. When a hungry newt glides into view, the larvae respond by wriggling toward the bottom. In spite of their sudden and confusing escape, the aquatic salamander gobbles scores of the squirming insects.

Deep in the spotted newt's digestive system, chemicals reduce the meal to manageable nutrients which fortify her ripening eggs. When these are ready, she embraces a mate to fertilize her eggs. Each egg is fastened individually to aquatic plants where the egg cells divide rapidly within their gelatinous sheath. The dark ball of cells gradually changes into a crescent shape. About 25 days later, a fully aquatic larva emerges. Frilly gills sprout from the neck area to absorb oxygen. A flattened tail pushes the greenish-yellow newt through the shallow water.

As the larva grows, the external gills are absorbed and lungs develop. With the loss of gills, the newt crawls from the pond to spend two or three years in a woodland environment where it is known as a red elf.

The red elf hunts for insects under rotting leaves and decaying logs; it even ventures out in plain sight. Although the reddish skin makes it obvious, a red elf has little to fear from predators. A disagreeable chemical produced by glands beneath the skin repels the hungry hunters.

After reaching maturity, the red-spotted newt returns to the water. The tail flattens into a broad, powerful fin, and the body color shifts from red to yellow. A row of crimson spots lines each flank. In spite of the newt's return to an aquatic environment, the lungs are retained to link the two worlds of this amphibian.

As you sit quietly at lakeside after a day of exploring, life continues in the water. Beneath the surface, tiny plants and animals begin their downward migration. Microscopic predators, like the phantom midge larva and the hungry water flea,

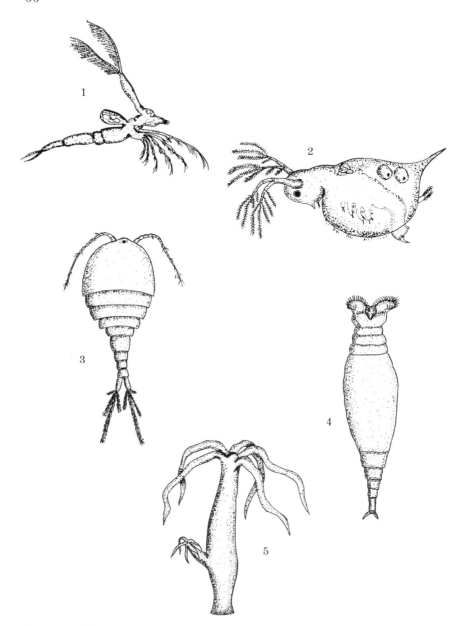

Some very bizzare creatures can be found suspended in between surface and bottom: (1) *Leptodora*, a predatory water flea, (2) *Daphnia*, a common plant-eating water flea, (3) *Cyclops*, a "one-eyed" crustacean, (4) a rotifer *(Philodina)* with a typical cilia-covered mouth, (5) *Hydra*, an animal with stinging tentacles.

The red elf stage of the spotted newt lives on land.

Leptodora, follow close behind in search of a meaty morsel. In contrast, phantom midge pupae rise to the surface where the pupal case splits and adult midges emerge. Floating on the empty cases, they dry their wings and join a dancing cloud of bugs above the water.

Perch and bass actively feed. They rely on their sense of smell and fine-tuned lateral line to provide a meal. Backswimmers and diving beetles leave the pond and seek other hunting grounds.

As the sunlight disappears and darkness prevails, activity between the surface and bottom slows, but does not stop. Instead, most of the pond's life is digesting the day's adventures and preparing for the ones to come.

A great blue heron stalks for eels in a shallow pond.

Visitors To The Shore

On the topographical map, Musquacook Lake is merely a blue splotch, one of several dabbles linked together like a short chain. Surrounded by low mountains and located deep in Maine's northwest forests, my wife, Cynthia, and I had decided this would be an appealing campsite. Yet, from the stern seat of our canoe, the lake looked larger and more formidable than the map had indicated.

At the head of the lake, a dark hump, resembling a blackened log, seemed no closer in spite of our twenty-minute paddle. Without binoculars, we could only speculate on its identity: a deer, a moose, a bear, or a blackened log. But, the possibility of closely approaching one of the larger mammals kept us paddling throughout the late afternoon.

We paddled quietly and drifted as much as possible until the black hulk began to take shape, in fact, two shapes—the bulky body of a cow moose and her stilt-legged calf.

Belly deep in the water, the hump-backed cow chewed on stringy stems and leaves which dangled from her mouth. Water trickled from the horse-like muzzle. Her dark brown eyes stared in our direction, but had a vacant look. Down went the head with a froth of bubbles streaming from her submerged hair. Thirty seconds later, her drenched face reappeared with a fresh bunch of weeds clenched between the flat teeth. Beneath her chin drooped a soggy "bell" resembling a long goatee. Its wet hairs were plastered together and were oozing water. From a distance, the animal had looked black, but she was really a rusty brown with dark brown shoulders and flanks. As she waded into shallower water, the legs appeared straw-colored below the knees.

She stopped to nibble on some scented water lilies and grunted to her gawky calf. The youngster ambled alongside and nuzzled her flank while a gentle breeze swirled between us. Suddenly, the cow glared in our direction as she caught our scent. With the hair erect along her hump, the cow tossed her head, snorted to her calf, and plodded ashore. Sheets of water cascaded from their belly fur as they hastened into a thicket of spruce and hemlock.

Moose (*Alces alces*) are the largest members of the deer family, a group which includes elk, deer, and caribou. Ranging from eight to ten feet in length, a mature cow weighs from 600 to 800 pounds, while a bull reaches 900 to 1,400 pounds.

Yet, it is not the sheer bulk of the male which people find noticeable, but the unusual antlers. Spanning six feet and weighing sixty pounds, the palmate antlers are not an ornament to attract a winsome mate. Instead, the hardened rack is a battle weapon to drive off rivals.

Each April, the antlers begin their growth as small bumps. During the following months, the antlers develop into a spongy tissue wrapped in "velvet." The sensitive velvet supplies numerous blood vessels to nourish the growing bone. By the end of August, the blood vessels dry up. The velvet dies and hangs in tatters from the hardened bone. When the bull rubs the antlers against a tree, the remaining velvet is stripped away to reveal the broad-bladed weapons.

During the early fall, the bull behaves like a lovesick fool, grunting and bellowing to interested cows, but fiercely defending his rights against all rivals.

Mating occurs by the end of October and the rutting season passes. The bull settles down to a normal life. The now useless antlers are discarded in mid-winter and are gnawed by mice and squirrels.

Eight months after the mating season, twenty-five pound calves are born. Generally, a cow has only one calf, but occasionally twins, and rarely, triplets. Completely helpless at birth, a calf strengthens rapidly. By the end of the first week, the youngster can run well and is gaining weight at the rate of two pounds daily. Within five months, it weighs 200 pounds.

The mother fiercely defends her calf until the following spring. Then, with the approach of a new birth, the yearling is driven away to fend for itself.

After the moose wandered from sight, we decided to beach the canoe and rest on the graveled shore. Flecks of mica sparkled among the water-worn granite stones. A scraggly spruce leaned over the water, its roots exposed by the eroding wash of waves. Sitting among the twisted roots, we relaxed in Nature's embrace.

A small bird fluttered weakly above the surface of the lake. The stiff wing-beats seemed inadequate for flight. A shrill piping announced the bird's arrival as it landed on a large rock and hopped to the beach. With head tilted, the "teeter-tail" bobbed up and down. The enchanting bird was a spotted sandpiper (*Actitus macularia*) in full breeding plumage. The white breast was spattered with brown spots, much like a painter's bib. The bird scuttered along the beach for several minutes, thrusting its needle-like beak at insects hidden among the stones. Piping its shrill call, the sandpiper fluttered away when a blue-gray bird swooped over the beach.

The rattling call of the belted kingfisher (*Megaceryle alcyon*) echoed along the shore as the crested hunter glided into the tipped spruce. A broad band of rusty plumage across the white belly identified the female kingfisher. From her high perch, keen eyes scanned the surface for the tell-tale ripple of a fish. A dimple on the water betrayed a minnow's location fifty feet away. The kingfisher flashed into the air and hovered above the spot for several seconds. With wings half folded, she dived headlong into the water. When she surfaced, she held a wriggling shiner in her beak. Returning to her perch, the kingfisher slapped the fish senseless against the wood and swallowed the shiner headfirst.

The kingfisher was not the only bird which fished with great accuracy. Earlier in the day, we had observed a little green heron (*Butorides virescens*). The short-necked heron had been wading knee deep in a small feeder stream. Like his larger relative, the great blue heron, the little green had hunted with stealth and patience. The crested head had been hunched against the shoulders, drawing the dagger-like bill into a striking position. Each movement of the bright orange legs had been taken with great deliberation. Once established in a good fishing hole, the heron had waited patiently for many minutes before seizing a spotted newt. Remaining in the same spot, the little green had speared fish and frogs before abandoning the site to fish elsewhere.

Across the lake, an ancient white pine was gilded by the setting sun. The top of the tree had died decades ago and most of the limbs had broken off. But, the pine had not out-lived its usefulness. A great jumble of sticks, at least four feet in height and a hundred pounds in weight, was cradled among the remaining branches. Atop this massive nest, the flightless wings of a young osprey (*Pandion haliaetus*) were out-stretched against a slight breeze. The osprey nest was the accumulation of several generations' work. Each spring, the male had added

A large bull moose displays his velvet-covered antlers as he stands at the shore of a pond in August.

more branches. Some had been gathered on the ground, while others had been broken from trees by hard dives. His mate had lined the nest with mosses and grasses before shaping a shallow bowl for egg laying.

By early May, the osprey pair had begun to incubate two brown-splotched eggs. Thirty-five days later, the hatchlings had chipped through the egg shell with a temporary egg tooth to enter the world in a tan-striped coat of down.

The adult birds fed their offspring strips of flesh which had been torn from fish caught in the lake. By soaring high above the lake, the osprey had spied fish near the surface. With wings partly folded, a forceful dive had carried the osprey beneath the water. Just before entering the water, the feet had been extended, and powerful talons had seized the fish. With the meat clutched tightly, the osprey had flapped free of the water and carried the dying fish to a feeding tree. There, he had devoured part of the fish before sharing the catch with the young birds.

Before paddling back to camp, we walked along the shoreline to a small sandy spit. Here, the erosive waves had abrated the rocks into specks of quartz, feldspar, and black tourmaline. Chunks of dri-ki littered the beach. Beneath these pieces of bleached wood, numerous worms, rotifers, and protozoans lived, shielded from the drying effects of the sun.

Concentric ridges ringing the sandbar marked the various

water levels. The soft sand was crisscrossed with animal tracks, reminders of the lake's visitors.

The tiny tracks of the water shrew (*Sorex palustris*) were sprinkled at the water's edge. The small mammal had dug several shallow pits in search of tiger beetle larvae. The grotesque larvae burrowed vertical chambers in the sand which they used as hunting blinds. The larvae hanged in the burrows with their flattened heads at the entrance. When insects scurried too near, the tiger beetle larvae seized their prey with hooked jaws. But, in spite of their powerful jaws, the larvae had merely been another crunchy meal for the hungry shrew.

The water shrew ate any aquatic insects, small fish, and eggs. Occasionally, plants were also eaten. The shrew's hind feet were adapted for swimming with a special fringe of hair. When diving, the blackish-gray fur trapped a shimmering coat of air bubbles which helped the animal bob to the surface quickly.

The pointed tracks of white-tailed deer (*Odocoileus virginianus*) were pressed deeply into the sand. The deer had stopped to drink and to browse on the spatterdock growing near the shore.

The most common tracks on the beach were the hand-like prints of the raccoon (*Procyon lotor*). The tracks wove an erratic pattern across the beach. They entered the water in several

The white-tailed deer frequent the shores of lakes in search of spatterdock.

places, only to come ashore a few feet away. Bits of crayfish legs and empty mussel shells were clues to the nocturnal hunter's menu.

As we paddled back to camp, the shoreline became two-dimensional in the fading light. Somewhere among the silhouetted evergreens, a chilling cry echoed along the shore. Unlike the moody tremolo of the loon which enriched the lake with mystery, the "quavering whistle" of the screech owl had a ghoulish tone. The owl was calling to announce his claim to his lake-front property.

Above our heads, fluttering wings could be heard as tiny bats flew erratically about the sky. We watched their silhouettes dip and soar in rapid pursuit of nocturnal insects which they caught acrobatically. Although the tiny mammals flew too swiftly for detailed observation, they were capturing insects in an apron of flesh stretched between their hind legs. Scooped into this "catcher's mitt," the insect was quickly devoured as the bat tucked his face into the apron and did a complete somersault.

Finding some pebbles in the bottom of the canoe, I tossed them into the flight paths of approaching bats. They responded instantly and darted toward the rocks, only to swerve away at the last moment. The bats were locating the pebbles and insects by echo-sounding. A steady stream of high-pitched squeaks, too high for people to hear, was being emitted by the hunting bats. When the sound waves struck an insect, they bounced back to the tall, pointed ears, telling the bat where the meal was located. Detecting their meat by sonar, the bats fed until stuffed.

By the time our canoe scraped the gravel shore, darkness had firmly enveloped our campsite. Weary muscles communicated their resentment of the busy day in unmistakable messages. Gratefully, we had pitched our tent before launching the canoe, and only supper preparation remained. The veil of stars in the Milky Way arched across the night sky, and the familiar constellations of Hercules, Cassiopela, and the Dippers shone more brightly than at home. Here, the interfering glow of street lamps and shopping malls did not diminish the ancient starlight.

As we relaxed near our campfire, a chorus of night sounds drifted to our ears. A wood peewee sang his final melody of the evening, fish rippled the lake's surface, crickets sawed a love song on their toothed wings, and the bass-voiced bullfrogs "*chug-o-rummed*" near the shore. After crawling into our sleeping bags, these same sounds lulled us to sleep.

I am still not certain what stirred me from that first night's slumber. Perhaps it was the faint scratching sound some distance away, or maybe the heavy silence along the shore which had been so noisy. But as I awoke, a rythmic shuffling whispered from the grass behind our tent. The faint padding paused, followed by a barely detectable sniffing. Then the soft scuffing resumed and stopped at the back of our tent.

As I sat in the heavy blackness of our small tent, innumerable stories from poorly informed "woodsmen" flooded my mind. A fierce pounding vibrated in my ears and throat as adrenalin accelerated my heartbeat and tensed my muscles. I froze as a sniffing nose was thrust against the tent wall and under the flimsy tent floor, followed by several deep-throated snorts. Immediately, the curious snout withdrew. My first thought was of a black bear seeking a morsel in our tent, but all food had been removed. Then as the intruder moved beside the tent, he scraped his back against the tent ropes. A flood of relief swept through me because no bear could ever fit beneath those ropes which were only a foot above the ground.

For the first time, I wondered if my wife had heard the snorting. Gently reaching across her pillow, I was surprised when I did not touch her face. She had indeed heard the commotion and had assumed a fetal position deep in her sleeping bag.

Before I could begin convincing her that our ordeal was over, a rattling sound echoed in front of our tent. Anxiously pulling back the tent's storm flap, we played the flashlight about the campsite. Our intruder was busy inspecting a nearby garbage can with only a fat rump and a ringed tail protruding above the rim. Head down, the raccoon fished about the barrel for whatever tidbits were left behind.

Laughing at our imagined fears, my wife and I retired and slept soundly the rest of the night. In the morning, raccoon tracks were scattered on the sand. But, overlying the hand-like prints were large six-inch bear tracks! We had slept through the big mammal's visit.

What Could Be Purer?

"And who art thou?" said I to the soft falling shower,
which, strange to tell, gave me an answer, as here
translated:
"I am the Poem of Earth," said the voice of rain,
"Eternal I rise, impalpable out of the land, and the
     bottomless sea,
Upward to heaven, whence, vaguely form'd, altogether
     changed, and yet the same,
I descend to lave the drouths, atomies, dust layers of
     the globe,
And all that in them without me were seeds only, latent,
     unborn;
And forever, by day and night, I give back life to my
     own origin and make pure and beautify it;"

"The Voice of the Rain"
by
Walt Whitman

What could be purer than fresh fallen rain? Glistening from
each blade of grass, dripping from every maple leaf, the rain is
cleansed by solar distillation. Evaporated from the land and
Earth's vast reservoirs, transpired through the pores of green
leaves, its vapor rises into the sky by the trillions of tons each
day. Left behind are the salts, minerals, and pollutants accumulated during its stay on Earth. Taken away is a bit of solar
energy, enough to excite its molecules to change from liquid into
gas.

Cooled and condensed around a tiny speck of dust, a mist of
droplets joins forces and troops together as clouds. Buffeted by
winds and thermals, the fleecy wanderers may travel hundreds
of miles or only a few. Finally, droplets unite into drops, and the
trillions of tons of water return as rain, snow, sleet, and hail to
sustain the biosphere's vitality.

This vision of precipitation's importance eluded my fishing

companion, Sam, and myself as we huddled beneath the stout boughs of a white pine. Cold drops splattered on the narrow strip of sand which rimmed the pond, and a fine mist squeezed through the tassels of pine needles. We hunched deeper into our ponchos and thought of the fat brook trout waiting to be caught. It mattered not to them that rain was falling.

A short distance from where we sat, a female wood duck called to her mate. Within moments, the drake splashed into the water beside their nesting house. The female dropped from the oval opening and joined her mate. Deep in the wooden box, her clutch of eggs was covered with warm down.

Numerous nesting boxes had been erected along the shore, most on wooden poles sheathed with broad bands of aluminum. The metal prevented hungry raccoons from shinning up the poles and stealing an easy meal of wood duck eggs.

Favorable nesting sites are equally important to the brook trout, lake trout, and landlocked salmon prowling the cold water of a lake. Without adequate spawning areas, their populations diminish. Natural reproduction of the brook trout in most ponds is dependent upon cold, unpolluted streams. In the early fall, brookies migrate into the fast flowing riffles to build redds (nests). Turning sideways and flapping her tail, the female trout loosens stones and sediments which currents carry away, leaving a shallow pit in the bottom. Following courting, she releases her eggs into the depression as the male fertilizes them. Moving upstream, the female digs another redd and its sediments settle on the preivous nest and eggs.

Although trout and salmon spawn during the fall, bass, pickerel, and white perch prefer May and June. Often the nest of a bass can be seen in shallow water with the male protecting the eggs. In spite of his best efforts, the bass' nests and those of other fish can be easily damaged by curious people who in their eagerness to observe a fascinating event inadvertently destroy the very thing they wish to watch.

While pollution and destruction of spawning areas presents a threat to populations of fish, another danger lies in the introduction of fish not native to a particular region. When the carp was introduced into the United States from Asia during the early 1800s, backers were anticipating an economic bonanza; instead, an ecological disaster occurred. They became the "starling" of the fish world. Carp competed directly with native species and altered water conditions by stirring up bottom sediments. The clouded water reduced light penetration which lowered plant growth. This, in turn, affected links in the food chains from algae to ducks.

In addition to displacing native fish, introduced species frequently carry diseases and parasites which can quickly become uncontrollable where no natural resistance exists. Intense scientific study and evaluation is needed before introducing exotic fish to any lake. Where populations of native fish already exist, interbreeding may occur to produce hybrid fish from two distinct species. By definition, a species cannot breed successfully with any other species, but enough exceptions exist to make this questionable. The coydog (coyote-dog cross) and the splake (brook trout-lake trout cross) are two examples. Thus, hybridization is a subtle process which eliminates species by gene contamination.

A paradox exists in the realm of the duck hunter where law-abiding people create a serious problem for waterfowl, frequently without realizing it. Whenever a burst of pellets is fired at a duck, numerous bits of lead fall into the water and marshy borders of the lake. The seemingly insignificant pellets when massed together total more than six million pounds of lead annually in the United States alone.

Dabbling ducks probing the bottom for food and grit ingest the lead shot. Once swallowed, the poisonous metal is held in a grinding organ, the gizzard. Lead is worn off the pellets, enters the bird's body systems, and paralyzes the gizzard. If the poisoning is severe enough, the paralyzed organ prevents swallowing, and the duck starves to death in three to four weeks.

Other toxins in the environment, although not packaged in lead spheres, are more deadly and wider reaching with their poisonous effects. In a nation which produces about two million different chemicals for innumerable purposes, it is inevitable that some will find their way into our lakes, ponds, and reservoirs. Some are extremely poisonous insecticides and herbicides, while others are fertilizers and sewage.

For more than twenty years, forests, marshes, and swamps were sprayed with solutions of DDT to destroy insects like gypsy moths, spruce budworms, and mosquitoes. Tons of this organochlorine saturated the land and entered bodies of fresh water, either directly or by runoff from forest and croplands.

The effects of DDT on populations of animals varied considerably. Not only did the insecticide kill wildlife outright, but it removed various links from food chains. Once an organism's food source was destroyed, starvation occurred in those animals

Each year, thousands of dabbling waterfowl, like this female mallard, suffer and die from lead shot poisioning.

dependent upon that particular link. The effects were then magnified along the chain.

In other situations, the chemical did not kill but was absorbed into the tissues in low amounts. Although DDT was nearly insoluble in water, it quickly dissolved into the fatty tissues of animals. Microscopic crustaceans absorbed small quantities of DDT from the oily droplets of diatoms. The more diatoms they ate, the greater the dose. A minnow dining on crustaceans concentrated the DDT to a higher level. Smallmouth bass accumulated even more poison and passed a lethal dose to fish-eating birds like grebes and ospreys.

The effects of DDT on breeding animals were equally disastrous. Predatory birds like the osprey and bald eagles suffered declining populations from eggshell thinning and inviable embryoes. Peregrine falcons which fed on fish-eating birds showed similar problems. A link was also found between the concentration of DDT in the oil of lake trout eggs and the mortality of young trout.

Today, DDT and numerous other insecticides are banned in the United States because of their uncontrollable, destructive potential.

The rain had stopped, but a constant pattering of heavy drops continued to trickle from the canopy of leaves. The late afternoon sun burned away the gray clouds and warmed the damp air with shafts of amber light. Far from shore, dimples spread across the lake's surface as trout slurped insects from the thin film.

Sam was already fifteen yards from shore casting his line in graceful arcs. Anxiously, I slipped my rod together and threaded the line along the tapered pole. Next came a moment of decision: which fly should be tied to the transparent leader? Light Cahills, Royal Coachmen, Parmachene Belles, and Black Gnats reclined in my palm-sized fly box. Hooked forgeries, these dry flies and several others were tempting mimics to cast upon the smooth lake. The choice made, I slogged toward my favorite fishing hole.

The soft light glinted on round stones scattered across the bottom. The clear water snugged the waders tight against my thighs. The memories of the two-mile tromp through the alder swamp, of the stinging branches flicking my face, of the mud clutching my boots, and of the hours spent beneath the dripping pine faded quickly. Behind me, a small cove cut a deep curve from the arching shoreline. A beaver's lodge lay at the cove's edge. Fresh cuttings provided evidence of their continued work. A pair of goldeneyes flashed overhead, their stubbly wings whistling in the damp evening air.

Shadows stretched farther onto the mirrored lake where swarms of tiny midges fluttered above the surface. An insect hatch was underway. Numerous midge pupae bobbed to the surface. When they touched the film, the rice-sized kernels split open and gray midges stretched from the pupal skins. Legs, bodies, and wings dried rapidly as the midges floated on their abandoned husks. The transformation completed, the midges slipped into the air on iridescent wings.

Drawn by the rising pupae, brook trout fanned along the cobbled bottom. Their sensitive eyes detected every movement. Flashing red and silver from their sleek flanks, the trout swirled to the surface to swallow hatching midges.

Prompted by the rising trout, I secured the fly to the tapered leader and casted toward the closest action. Before the fly touched the water, a brookie greedily snatched the replica. The line tightened, the rod bowed, and the water churned as the determined trout fought courageously before coming to net. Wetting my hands, I slipped the muscled body from the mess, carefully removed the hook from the lower jaw, and returned the valiant fighter to the water. He lolled briefly near the surface, regained his bearings, and blended into the dark bottom. He would live to fight again.

Not all released fish are as fortunate. Statistics have shown that fish deeply hooked in the throat or stomach die within a day or two if the hook is removed. Surprisingly, if the hook is left in place and the line is cut, the fish has a greater chance of surviving. The hook will rust away. In addition, studies on flying fishing versus worm-angling indicate a much higher death rate among worm-hooked fish. Although this may vary with individual fishermen, a worm-baited hook is usually set deeper in the throat or stomach where it punctures a vital organ. A fly generally hooks the mouth or jaw.

Against hook and line, brook trout have a fighting chance. Their stamina, leaps, and instincts insure survival. But, what chance do they have against an invisible threat cloaked in their own element, the water?

A thousand miles west of this isolated lake, smoke stacks, some a hundred stories tall, are spewing sulfur dioxide, nitrogen oxides, and other pollutants into the air. These products result from the burning of fossil fuels, especially soft coal. Fed into the circulating currents of the atmosphere, these reactants

Proper angling techniques can reduce the number of fish killed needlessly.

drift hundreds, perhaps thousands, of miles. During their journey, chemical changes unite the deadly cargo with water, oxygen, and sunlight. Eventually, they condense into droplets of rain or flakes of snow and precipitate back to Earth as dilute solutions of sulfuric and nitric acid.

The effects of acid rain on forest and cropland are difficult to determine, but the consequences to freshwater lakes and ponds are conclusive. Deep in the Adirondack Mountains of New York, more than two hundred lakes and ponds are devoid of life. These bodies of water lie strangled by increasing acidity. Hundreds more are endangered throughout the Adirondacks and in nearby Canada. Nor is this problem confined to North America. Scandinavia, deluged by acidic pollutants from European industrial centers, has nearly 20,000 lakes and ponds whose crystalline blue waters are depleted of all fish, salamanders, frogs, clams, plankton, and numerous insects. No otters dart after fish. No muskrats crunch freshwater mussels. All that remains are acid-tolerant insects like whirligig beetles and water boatmen. Larger plants, intolerant of the acidic fluid, decline, while blue-green algae and sphagnum mosses thrive. In the bog-like environment, nutrient-recycling bacteria disappear and layers of peat may develop in the shallows. Although the lake is not completely "dead," link after link in the food chains have dissolved, drastically reducing the profusion of species to a few resistant specimens.

When Walt Whitman wrote of the rain in 1885, the Industrial Revolution was a developing child, hungry for coal; however, the amount of coal burned was small compared to our present energy glut. Yet, even then, acidity tainted the vital liquid. As early as 1852, a perceptive English scientist, Robert Angus Smith, had discovered acidity in rainwater. Like many scientists ahead of their time, his discovery was ignored for a hundred years.

Today, power generating companies and heavy industries in the United States expel 26 million tons of sulfur dioxide and another 22 million tons of nitrogen oxides yearly into our nation's atmosphere. Acid rain can no longer be ignored.

Intensive research is being conducted to find ways to reduce emissions which cause the rain and to neutralize the acid already present in our lakes. One question which can now be answered is, "How acid is acid rain?"

Acidity is measured on the pH scale which ranges from 0 to 14. A pH of 7 on this scale is given to distilled water, a neutral compound. Acidity is attributed to the numbers less than 7, alkalinity to those greater than 7. Thus, a chemical with a pH of 5 is more acidic than another chemical at pH 6. In fact, it is ten times more acidic since each unit on the scale represents a ten-fold change.

Normal rain with a pH of 5.6 is always slightly acidic because of its reaction with the atmospheric gas, carbon-dioxide. Glacial records of "fossil rain" indicate a pH generally greater than 5 before the Industrial Revolution. Precipitation today over the eastern United States ranges from pH 4.5 to 4.0.

As acid rain patters onto the ground around a lake, it seeps through soil and removes various metals. Several of these dissolved metals, aluminum, mercury, and lead are easily carried to lakes by runoff and are toxic to life. Young fish, for example, collect aluminum in their gills where large amounts of mucus are produced to counteract the poison. Instead of helping, the mucus clogs the gills and suffocates the fish.

Research has also demonstrated that each lake may react to acidity in different ways. Lakes carved from layers of limestone and sandstone can neutralize or buffer the acid through chemical reactions, while those gouged from solid granite have no buffering effect. Varying degrees of buffering by the lakes between these extremes account for the wide variation in lake pH.

Taking advantage of the neutralizing effect of lime, both Sweden and New York have dumped tons of calcium carbonate into selected lakes as a stopgap measure to preserve fish populations. This procedure, however, is expensive and the effects are short-term.

Surely what is needed is to control the pollutants before they enter the atmosphere, through effective legislation, burning low-sulfur coal, and installing "scrubbers" on existing power plants to reduce emissions. The cost of this process, estimated to be in the billions of dollars, is an expensive price tag which will ultimately be borne by consumers. Acid rain like so many environmental problems has no easy answers, only bitter solutions, but the price of ignoring the dilemma will be incalculably greater.

Winter, Spring, Summer, Fall

December twenty-second dawns crisp and clear. Wisps of cirrus clouds streak the eastern sky with gold and red. On this day of the winter solstice, the sun seems to pause briefly as it reaches its greatest distance from the celestial equator before renewing its northern course.

As the winter sun swells above a shaggy silhouette of white pines, crimson rays dabble highlights on the bark of trees and shrubs. The white breast of a chickadee blushes in the morning glow. Hanging upside-down on a spruce bough, the black-capped bird thumps a cone with his stubby beak to loosen any remaining seeds. These seeds, like millions of others, are time capsules of life awaiting the warmth of spring to stir their growth.

Beneath the insulating ice and snow covering the pond, similar bits of sleeping protoplasm lie protected in the mud and leaves. Seeds of spatterdock, pickerel weed, and pipewort are safely tucked away along with the "seeds" of many animals.

Confronted with a frigid environment which can become a solid chunk of ice, some residents take evasive action and migrate to warmer climates. Bats, ducks, geese, and even insects escape the freezing temperatures by flying south.

Several animals escape an icy death by altering their body functions. Many frogs bury themselves in the muddy bottom where temperatures hover slightly above freezing. As the bodies of these cold-blooded amphibians assume the temperature of the surrounding mud, they become less and less active. Their heart rate drops dramatically, and they absorb oxygen through their moist skin. In this state of hibernation, the energy needed to sustain life comes from a reservoir of yellow fat.

Sharing the muddy bed are turtles and numerous insects like *Dystiscus*, back swimmers, water striders, and water boatman. Their inactive sleep requires only a trace of oxygen.

Sunlight radiating through the ice permits photosynthesis to continue. This replenishes valuable supplies of oxygen. Without it, many creatures would suffocate. Larger lakes generally do not have oxygen depletion, except near the bottom where bacteria continue to rot dead organisms.

Cold-blooded fish, their body chemistry slowed in the chilled water, move sluggishly. Their appetite slackens, but feeding does continue, as any ice fisherman can tell you. Dangling a baited hook through a hole in the ice often produces surprising results to those unaccustomed to winter fishing.

Otters actively fish all winter. Entering the water through a submerged tunnel, they prowl beneath the ice and snatch the sluggish fish with ease. Air bubbles trapped under the ice provide otters with occasional gulps of oxygen.

Beavers and muskrats also swim beneath the ice in search of food. Beavers chew the bark from the supply of aspen boughs which they have weighted to the bottom with rocks. Muskrats poke along the mud for roots and mussels. Returning to their thatched dens, muskrats rest above water level on beds of dry grasses. A muskrat house, woven from sedges, rushes, and cattails, can also be eaten if other food is unavailable.

Although life in the winter pond is reduced to its lowest ebb, evidence exists to indicate a change. The eggs of fairy shrimps hatch in the cold of January to produce tiny embryos. These change into multilegged adults which cruise about on their backs beneath the ice. The mother otter gives birth in February; she cradles and nurses her pups against her warm belly fur. Mayfly nymphs and other aquatic insects begin the changes required to reach adulthood.

Slender, silver-green fish called smelts *(Osmerus mordax)* swim into a lake's tributaries during late February and early March to begin spawning. As nightfall arrives, breeding groups of adult fish swirl together. The females release sticky eggs which are fertilized as they float freely in the water. The eggs adhere to rocks, sticks, even fish, and begin development immediately.

The fairy shrimp, the smelts, and the otters are all responding to a biological clock. This internal mechanism is set by the increasing daylight of late winter and signals the warm days ahead when the pond's life will blossom once again.

— — — — — —

Once the crackling skin of winter ice has melted from a pond, thumbnail-sized tree frogs appear in the water to seek mates. At first, they remain silent. Only the dried cattails rattle in the cool wind. Each day, the sun hangs higher in the sky and warms the water, until one evening in early April a solo performance signals the spring peepers' return. Succeeding nights bring duets and trios of soprano-voiced bachelors. A symphony of trills, snores, gargles, and grunts follows as peepers, pickerel frogs, wood frogs, and leopard frogs all join in a spring concert.

Warming days of early spring effect many changes in lakes and ponds. During winter, a thick canopy of ice has kept the water cold. No oxygen has entered. Deep in lakes, bacteria have continued to slowly decay dead plants, removing oxygen and releasing nutrients.

As the water warms, the changing temperatures create currents. Combined with wind, an "overturn" mixes the water and blends the temperature, oxygen, and nutrients evenly throughout the lake.

The surge of nutrients into the water fertilizes the growth of diatoms and desmids. These microscopic plants "bloom" into millions of living bits of protoplasm, actively changing sunlight to sugar. With high energy food to fuel their activities, their numbers continue to rise, but not unchecked. *Daphnia*, rotifers, and copepods hatch daily and devour the waiting harvest.

Green strings of algae lengthen in the shallow water. Their photosynthesis produces bubbles of oxygen which buoy the slender fronds to the surface. Oxygen dissolved from these bubbles supplies the gills of freshly-awakened creatures. A dragonfly nymph lies in ambush among the plant stems. A caddisfly larva, secure in his home of cemented twigs and bits of leaves, pokes along the bottom. Stringy nematod worms throw their hair-thin bodies into S-shaped kinks. The giant water bug hangs motionless on a cattail stalk.

During early spring, bass and sunfish move from the deep water where they spent the winter to the warmer, foodrich shallows. Pickerel splash among the marshy borders as mating occurs and their adhesive eggs are dropped to the bottom.

In the damp earth beside the pond, bright green whorls of skunk cabbage leaves unfold. The mottled red horns of this plant produce an offensive odor which quickly attracts pollinating flys. Toads, hibernating in these borders, push to the surface and hop toward the pond. Salamanders, too, are drawn to the water where they must return to lay their eggs.

Flocks of redwinged blackbirds sweep into the marshy pond borders. Flashing the red badge of their species, they teeter among the cattails and feather grass. When a muskrat rattles through the cattails, the blackbirds burst into flight with a whir of wings.

Ducks arrive daily to dabble and dive. Blacks, mallards, mergansers, ringnecks, and wood ducks rest and feed. Many push north, but others remain to nest among the weeds and in hollowed trees.

Green shoots appear at the surface as perennial roots and tubers of emergent plants push their sprouts skyward. Cattails and rushes, pickerel weed and arrowhead stretch toward the sun. The tiny leaves of duckweed sprinkle the surface with dashes of green.

With the arrival of May's warmth, lakes and ponds explode with life. Not only are animals trying to reach the pond, many want to leave. Dragonfly nymphs climb from the water on plant stems, spit their yellow-brown skin down the back, and emerge as winged adults. Diving beetles fly from pond to pond and stop where hunting is best.

After a year underwater, mayfly nymphs emerge to shed their outer skin. Gradually, they draw their fragile bodies from the crusty husk to expose delicately veined wings. Not yet ready for adult life, they shed a second transparent skin. When their bodies have dried, these non-feeding adult mayflies rise into the air to mate and die. Most live only a few hours as adults.

Sunfish, suckers, and bullheads begin their spawning activities during the late spring. Sunfish can be seen as they guard their nests in shallow water.

Spring is also the time of renewal for the American bittern. Nicknamed the "thunder pumper" for its hollow pumping call, this heron is streaked with brown and white, a camouflage perfectly adapted for the marshy edges of a lake. Not only does the plumage blend with the plant leaves, but the bittern pokes his bayonette-like beak skyward and sways in rhythm with the wind-blown weeds. The bittern's young, while still in the nest, also teeter instinctively when danger threatens. However, the deception is less effective than their parents' since they are frequently out of sequence with the bending plants.

Gradually, this explosion of new life dwindles as harried parents search for food to fill the hungry mouths. Spring, the time of birth, passes for another year, to be replaced by the hot months of summer.

— — — — — —

Over-wintered eggs have hatched. The fragrant water lilies have spread their broad green sugar factories across the shallows. A steady hum of insects rises from the weeds along the shore. Birds' melodies of courtship are replaced by the demanding peeps of hatchlings with empty stomachs. Mammals born in the spring venture forth with increasing regularity and independence. Summer is a time of growth and development, a time to learn survival skills and behavior patterns which will carry them through the hazzards of pond life.

The female diving beetle, *Dystiscus*, holds her mate securely and cements a cluster of eggs to his back. Carried during development, the eggs are protected from numerous predators, including the male. His voracious appetite respects no family ties. He would quickly devour his own unhatched children if he could reach their dorsal cradle.

For nearly a week, remarkable changes inside the egg alter the original cell into a ball of specialized structures. Eyes, heart, nerves, jaws, and stomach all differentiate in the growing embryo. Although the embryos are protected from their gluttonous parent, they are far from safe. As the tiny water tigers begin to hatch, the first born impales a sibling with his sickle-shaped jaws and drains the body fluids.

Along the damp borders of a pond, the rich earth houses thousands of earthworms. They constantly tunnel through the dirt, swallow the soil, and digest the bits of plant material surrounding each particle. These plump annelids attract many hungry birds, including the Wilson's snipe *(Capella gallinage)*. Using a long slender beak, the snipe probes the soft earth. Striking a worm, the beak's flexible tip opens to grip the prey and pull it from the ground. When sitting in the nest, the marvelous arrangement of brown and white feathers, of dusky spots and black bars, enables the snipe to vanish into its surroundings.

Muskrats enjoy the warm waters of summer. Much time is spent gathering cattails and sedges to build their summer houses. Food is plentiful, and cattails, rushes, blue flags, and arrowheads complete the salad menu. For meat, freshwater mussels are snatched from the mud and carried to feeding platforms of matted weeds. The mollusks are nimbly shucked, eaten, and the shells cast onto a growing pile nearby.

Sunbathing is a favorite pastime of the pond's reptiles. Turtles, in particular, crawl out on rocks or logs and bask in the warm sun. The heat they collect increases body activities and quickens cold muscles. Sunning also dries up algae growing on the shell and discourages parasites such as leeches.

During the summer, lacy strands of plant-like growths stretch around the stems of pickerel weed and spatterdock. Some resemble miniature ferns. A few grow into flimsy, gleatinous balls which quake with every disturbance in the water. These organisms do not look like animals; yet, under magnification, transparent tentacles sweep the water for bits of food. Bryozoans, the moss animals, strain diatoms, algae, and protozoans into a central mouth. Each tentacled individual is part of a larger, constantly expanding, colony. Sensitive to water pollutants, bryozoans grow only in clean, fresh water.

As the sun continues to warm the banks of the pond, numerous wildflowers respond with bursts of color. The brilliant cardinal flower *(Lobelia cardinalis)* grows at the water's edge, its crimson blossoms clustered along the stem. An aquatic cousin, the water Lobelia *(Lobelia dortmanna)*, sprouts its pale blue flowers from a stem growing in the shallows. Asters add splashes of yellow, gentians and bell flowers clumps of blue, as the colors of summer are randomly scattered for our enjoyment. Look, but do not pick since many of these enchanting blooms are rare.

This same heat which warms the shore also raises the temperature of the water. Lakes develop distinct temperature layers. The warm surface water is separated from the cold bottom layer by a boundary of rapidly cooling water, the thermocline. The thermocline is an effective barrier to prevent mixing of nutrients, temperatures, and oxygen. Cold-water fish, like trout and salmon, hunt below the thermocline, while bass and pickerel lurk beneath the shade of rocks and plants in warmer water.

Fish living in ponds must be very tolerant of the summer heat since the shallow water warms from top to bottom. Animals found in small ponds face the added threat of drying. Fish trapped in evaporating ponds will die, but better equipped animals can adapt. Adult insects simply migrate to another pond along with most frogs and turtles. Snails, clams, *Daphnia*, and leeches enter a state of dormancy, called estivation. Estivating animals have lowered body functions and needs, allowing them to survive in the sun-cracked mud. Protozoans and bacteria grow a protective cyst. Their lightweight cases can protect the bits of protoplasm for several years, or they can be blown on the wind to another pond. When the rain restores the pond, these sleeping creatures will awaken to repopulate the water.

The warm summer water attracts many visitors. Raccoons,

flipping rocks and sticks in search of crayfish, scurry along the shore. Moose browse on water lilies and tender shoots. By far, however, the most numerous visitors are people. Ponds and lakes are areas of limitless fascination at any time of the day or night. When treated with respect, they also provide opportunities for fishing, swimming, snorkelling, canoeing, and camping.

— — — — — —

Snow can come surprisingly early some years. One September I remember, the trees on the flanks of Bigelow Mountain were still more green than yellow. Specks of red, gold, orange, and russet marked the leaves like tiny stained-glass windows. A rime of crystals edged every blade and stalk. Powdery flakes clung to the spruce needles and hid in the shadows of the gnarled bark. Cinnamon caps of a few mushrooms pushed through scarlet mats of frost-touched sphagnums.

Cynthia and I stood in the col between Cranberry Peak and the Horns, 2,500 feet above Flagstaff Lake. At this elevation, the leaves already displayed their autumn colors. The birches were brushed with tarnished gold. Deeply toothed maples glowed in scarlet; the lightly serrated beech in copper. Clusters of golden needles splayed from the boughs of a tamarack.

Cranberry Pond, a mountain tarn, lay clear and cold before us. Water-worn rocks were strewn across its bottom and along the shore. A jumble of colorful leaves littered the stones and drifted on the surface. Small fish rummaged for insects among the soggy leaves which had settled to the bottom. These leaves would slowly rot, returning nutrients to the pond to help fertilize the growth of diatoms and algae.

We ate our lunch in warm sunlight and heavy silence. The sounds of spring and summer, of singing birds, croaking frogs and humming bugs, were gone. Stilled by the frost and shortening days, their internal clocks had warned them of the approaching winter.

Many insects had already died or migrated south. Countless others were tucked beneath the mud and rocks at the bottom of the pond. Some over-wintered as adults, some as larvae. Others spawned resistant eggs.

Bryozoans and sponges died in the chilled water and fell apart, but their tough-coated winter buds slept in the sediments with the eggs of rotifers and fairy shrimps.

Frogs and turtles were buried in the bottom ooze, while toads and salamanders slept in secure pockets beneath the forest litter and rotting logs.

Gone too were the broad leaves and fragrant blooms of the pond lilies, but deep in the mud their starchy roots stored the energy and the life force required to restore the plants. The seeds of innumerable water weeds joined the sleeping millions at the bottom along with the spores of algae. A few of the more hardy algae would remain active, photosynthesizing the rations of sunlight transmitted through the snow and ice.

As we finished our lunch, a pair of mallards paddled into the protective lee of our cove. For several minutes they dabbled bottoms-up, their orange feet thrusting against the water for leverage. When they had eaten their fill, the pair flapped from the pond in a flurry of sparkling spray and returned to their migratory escape.

Everything packed, we hiked down the trail. After descending a thousand feet, we rested on an outcropping of stone. The wooded flanks of Sugarloaf and Crocker Mountain stretched before us.

Behind us, a faint, far-away sound stirred. It was the familiar "ronk-a-ronk-ronk" of Canada geese on their southward flight. Their clamor grew louder until the V-formation swept low overhead. They were tracing the contours of the mountain and passed so close we could hear the air rushing through their feathers. Following the gander in the lead, the flock plunged a hundred feet below us, leveled off, and pointed toward Caribou Valley between Sugarloaf and Crocker. The honking faded, and the wedge of broad, brown wings slipped from view. Guided by mysterious clues and pushed by a northeast wind, "the forerunners of winter" would travel many miles before resting in a marsh late that night.

With the passing of the geese, the silence of our autumn day returned, a silence which would endure through the snows of winter, until the sounds of returning geese awakened the spring.